3

NEMERTEANS

Biological Sciences

Editor
PROFESSOR A. J. CAIN
M.A., D.PHIL
Professor of Zoology
in the University of Liverpool

NEMERTEANS

Ray Gibson

Lecturer in Marine Zoology,
Liverpool Polytechnic
and
Honorary Research Fellow,
Department of Zoology, Liverpool University

HUTCHINSON UNIVERSITY LIBRARY
LONDON

HUTCHINSON & CO (*Publishers*) LTD
3 Fitzroy Square, London W1

London Melbourne Sydney Auckland
Wellington Johannesburg Cape Town
and agencies throughout the world

First published 1972

*Jacket/cover picture shows typical nemertean colour
pattern superimposed on photograph of* Cephalothrix
swallowing prey

*This book has been set in Times type, printed in Great Britain
on antique wove paper by Anchor Press, and
bound by Wm. Brendon, both of Tiptree, Essex*

ISBN 0 09 111990 1 (cased)
0 09 111991 X (paper)

321 180

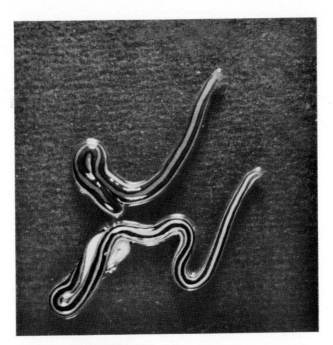

Above: Plate 1
Colour patterns of *Geonemertes
pantini* (above) and *Geonemertes
novaezealandiae* (below). Photograph by Dr Elizabeth J. Batham.

Left: Plate 2
The blood system of
Malacobdella grossa,
demonstrated by a histochemical
method for visualising
exopeptidase activity. Original
photograph by the author, first
published in black and white in
Gibson and Jennings (1967).

TO MY WIFE

CONTENTS

ACKNOWLEDGMENTS

I wish to express my appreciation to the many friends and colleagues whose assistance has greatly helped in the preparation of this book. In particular, grateful thanks are due to Professor A. J. Cain and Dr Janet Moore who have read and criticised the complete manuscript, to Dr Fumio Iwata and Dr John J. Poluhowich for commenting on certain chapters, to Miss Anne M. Fishburn for her invaluable aid in obtaining so many of the references, and to Mrs Jacqueline Hibbert for translating German. In addition Mr D. G. Eason has kindly allowed me to cite his unpublished results.

For kindly providing original photographs I am indebted to Dr Elizabeth J. Batham (Plate 1), Dr J. B. Jennings (Plate 3A) and Dr K. J. Pedersen (Plates 3B, 4A,B). For permission to reproduce other figures, redrawn or modified, I am grateful to Dr Stefano Bianchi, Dr Lise Brunberg-Nielsen, Dr Carl-Erik Cantell, Professor R. B. Clark, Professor Diva Diniz Corrêa, Professor H. Friedrich, Professor V. V. Hickman, Dr Fumio Iwata, Dr E.-A. Ling and Professor E. N. Willmer. Reference to their work is given in the appropriate figure captions. Dr Iwata has also kindly let me use an original unpublished drawing for Fig. 33.

For permission to reproduce copyright material I am grateful to the Editors of *Zoologischer Anzeiger* (Figs. 1A, 2L, 8F, 9F, 24D), *Zeitschrift wiss. Zoologie* (Figs. 10D, 14A, 15E), *Ophelia* (Figs. 8G,H), *Nature* (Fig. 9E), *Comparative Biochemistry and Physiology* (Plate 2, Fig. 14G), *Journal of Experimental Zoology* (Figs. 24A–C, 26A–F), *Biological Bulletin* (Figs. 6A–C, 7C–F, 8A,B, 10F, 11A,B, 15C,D,F, 25) and *Embryologia* (Figs. 27A,B, 32B,D); the Director of the Museum of Comparative Zoology, Harvard University (Figs. 1B,C, 2B,D,E,I–K, 3B–G, 4A,C,F, 7H, 8C,D,J, 9A–D, 12A,B,G, 13A,B, 14B–E, 15A,B, 16B,C,E–

G); the University of Illinois Press (Fig. 2G); the McGraw-Hill Book Company (Figs. 27C,D, 29G–J, 30A–C, 32A,C,E,F); the New York Zoological Society (Fig. 3A); the Company of Biologists (Figs. 4G, 13D–F,H, 19, 23); the Marine Biological Association of the United Kingdom (Fig. 6D); Akademische Verlagsgesellschaft, Geest und Portig K.-G., Leipzig (Figs. 10A,B, 12D, 13C); the Zoological Society of London (Fig. 10E); the Trustees of the British Museum (Natural History) (Figs. 11C,D, 12C, 14F, 15G,H); Academic Press, New York (Figs. 11E–H, 20, 22); the Royal Society of Tasmania (Figs. 29B–F); the Director of the Akkeshi Marine Biological Station (Figs. 28, 29K–M); the Royal Swedish Academy of Science (Figs. 28, 32C_1–C_3); the Zoologiska Institutionem, Uppsala Universitet (Fig. 17); the Academy of Natural Sciences (Figs. 4B,E); Springer-Verlag, Berlin, Heidelberg and New York (Plates 3B, 4A,B) and the Connecticut Academy of Arts and Sciences (Figs. 8E, 13G).

Finally, and not least of all, my thanks are due to my wife, Patricia, for her unfailing encouragement and help in so many ways.

RAY GIBSON

Liverpool

INTRODUCTION

Members of the phylum Rhynchocoela (= Nemertea, Nemertinea or Nemertini) are bilaterally symmetrical worms representing a peak in acoelomate development. They possess many morphological features in common with turbellarian platyhelminths, including a protonephridial excretory system, the presence of a solid parenchymatous tissue, and the absence of either respiratory or skeletal structures. Similarities also exist in the structure of the ocelli and certain other sensory organs, in cellular elements of the body wall, and in the spiral cleavage of the eggs.

The two major advancements shown by nemerteans are the development of a gut with separate mouth and anus, and the introduction of a parenchymatous blood vascular system. With the exception of a few unconventional species of trematodes, no organism below the nemertean level of organisation possesses either of these two features.

Two nemertean characteristics that are of particular interest from a phylogenetic aspect are the common occurrence of epidermal rhabdite-like structures, found also in the majority of turbellarian flatworms, and the possession of an eversible tubular proboscis, a somewhat similar organ being found additionally in kalyptorhynchian Turbellaria. The nemertean proboscis is borne within a dorsal fluid-filled cavity, the rhynchocoel, which, in the majority of species, extends for most of the body length.

Other morphological features found in the phylum are the well-developed nervous system, comprising paired cerebral ganglia and main lateral longitudinal nerves; the frequent internal pseudometamerism shown by both the gonads and the intestinal diverticula; and the presence of cilia and microvilli in both the epidermis and gut epithelium.

The phylum is a relatively small one, estimates on the number of species varying from about 550 (Hyman, 1951) up to 750 (Mayr, Linsley and Usinger, 1953). Many of the so-called species are known from only single or damaged specimens, but others are common, if unobtrusive, inhabitants of a variety of habitats. The vast majority of nemerteans are dioecious, free-living marine animals, dwelling beneath stones, burrowing into soft muddy or sandy substrates, living amongst algae or in rock crevices, or sheltering in the lee of various sessile invertebrates. Exceptional habits are found, however, in several, particularly enoplan, forms. Examples include species of *Coenemertes*, *Dichonemertes*, *Geonemertes*, *Potamonemertes* and *Prostoma* that are hermaphroditic; *Paralineus*, *Potamonemertes* and *Prostoma* that live in fresh water; *Geonemertes*, variously arboricolous, terrestrial or supralittoral; *Carcinonemertes*, parasitic on crabs; and the species that are commensal with either bivalve molluscs (*Malacobdella*) or cirripedes (*Nemertopsis*). Other species may be found as commensals of tunicates, living in the pharyngeal cavity (*Tetrastemma*) or the atrium (*Gononemertes*). Several rather exotic forms are bathypelagic, examples including *Chuniella*, *Pelagonemertes* and *Planktonemertes*, members of the genus *Phallonemertes* additionally being provided with simple copulatory organs developed from modified testes. Some species of *Geonemertes*, *Poikilonemertes* and *Prosorhochmus* possess the unusual feature of combining hermaphroditism with ovoviviparity.

Rhynchocoelan classification is based upon morphological structure, two principal divisions being recognised. The class Anopla, with the mouth positioned posterior to the cerebral ganglia and the proboscis without stylet armature, contains the Palaeonemertini[1] and the Heteronemertini, whilst the Enopla, with mouth anterior to the cerebral ganglia, comprises the Hoplonemertini, in which the proboscis is armed with one or more needle-like stylets, and the morphologically atypical commensal group, the Bdellonemertini.

Nemertean-like animals have existed since the Cambrian period at least (MacLeay, 1839), but early zoological authors were seemingly unfamiliar with them. Nemerteans were not reported in literature until Borlase (1758) described a 'Sea Long-Worm' as '. . . the long worm . . . I chuse to place here among the less perfect kind of sea animals . . .'. Borlase, like many authors following him, had no idea of the systematic position of his specimen, and numerous erroneous reports concerning the group were published during the next hundred

[1] Iwata (1960a) divides the palaeonemerteans into the two orders Archi- and Palaeonemertini on the basis of embryological evidence. His suggested reclassification is discussed in Chapter 1.

or so years. Amongst the confusion of the period nemerteans were variously described as marine 'insects' (Baster, 1762), annelids (Pallas, 1766; MacLeay, 1839), planarians (Müller, 1773, 1774; Fabricius, 1780; Linnaeus, 1788; Rathke, 1799; Lamarck, 1801), leeches (Bosc, 1802) or molluscs (Girard, 1851), and for many years remained wrongly associated with these invertebrate groups.

The majority of problems concerning nemertean anatomy and systematics seem to have resulted from difficulties in interpreting the functions of the various body components. The proboscis, for example, was commonly referred to as the alimentary canal (Pallas, 1766; Fabricius, 1780; Dugès, 1830; Johnston, 1837; Leidy, 1850, 1851; Williams, 1852; Baird, 1866), and presumably as a result of this misinterpretation in function the proboscis pore was for long regarded as a common mouth and anus. This error persisted even after Müller (1788–1806) and Fabricius (1797) correctly identified the position of the anus in anoplan and lineid forms respectively. Other, more unusual, roles attributed to the proboscis included that of a penis (Huschke, 1830; Leuckart, 1830; Örsted, 1844) and a tactile sensory organ (Delle Chiaje, 1825; Rathke, 1843). Several authors, although noting the presence of a mouth, wrongly ascribed to it the function of genital aperture (Ehrenberg, 1831; Renier, 1847; Girard, 1853; Stimpson, 1855; Johnston, 1865). Comparable errors regarding the function of such structures as the cephalic slits and grooves were equally common during this era of nemertean history.

Cuvier (1817) was the first author to employ the name *Nemertes* (Greek, a sea nymph, Mediterranean rather than oceanic, a daughter of Nereus and Doris) in designating Borlase's (1758) species as *Nemertes borlasii*, at the same time observing that it should be placed in a new order. He was apparently unaware that the generic name *Lineus* had been previously allocated to this species by Sowerby (1806). Cuvier was also the first to distinguish between planarians and nemerteans, grouping the former with acanthocephalans as 'Intestinaux parenchymateux', the latter with roundworms as 'Intestinaux cavitaires'. It was after this publication that the group became known as nemerteans or nemertines, although it was several years later before the name became universally accepted. Other names suggested but not subsequently adopted included Teretularia (Blainville, 1828), Annelosi Polici (Delle Chiaje, 1841), Cestoidina (Örsted, 1844), Miocoela (Quatrefages, 1849), Aplocoela (Blanchard, 1849) and Turbellaria Proctucha (Huxley, 1877).

The first accurate treatise on nemertean anatomy was that published by Max Schultze in 1851. It was he who originally appreciated the function of the proboscis as 'an aggressive organ', and differentiated between the armed and unarmed types. Schultze was also the first to mention nephridia in nemerteans, which he defined as Turbellaria with an anus and eversible proboscis under the name Nemertina or Rhynchocoela. In 1853 the same author proposed the division of the Nemertina into Enopla and Anopla on the basis of the presence or absence of proboscis stylets respectively. His classification was far from complete, but marked a significant step in the literature of the group. Schultze was clearly the earliest to understand nemertean anatomy with any degree of accuracy, and Hyman's (1951) comment that '. . . It appears most just to adopt the name employed by that zoologist . . . We therefore term the phylum Rhynchocoela or Nemertina after Schultze . . .' pays due tribute to his work.

During the next few decades many workers published articles on one or another aspect of nemertean biology, although several were far from accurate. The period from 1851 to about 1900 represented a 'clearing and sorting' era in nemertean literature that culminated in a reasonable and quite extensive understanding of the group. Notable authors of this time included McIntosh (*1867–1906*),[1] Hubrecht (*1874–87*), Bürger (*1888–1909*) and Bergendal (*1891–1903*). The many articles of Bürger were particularly detailed and painstakingly executed and, in a number of instances, extremely comprehensive.

Amongst those who have performed research on nemerteans one name in particular stands out both for the extent of his work and for the time span of his numerous publications. Much of our current knowledge on many aspects of nemertean biology is due to the labours of Wesley R. Coe, who in the period 1895–1959 produced more than sixty articles devoted to this invertebrate group. Coe may justifiably be regarded as the founder of modern nemertean investigations.

The early part of the twentieth century proved to be a time when nemertean investigations became concentrated very much more upon specific aspects of their biology, with the result that in some fields our current knowledge is quite extensive, whilst in others it is at best minimal. Authors of importance at this time included Dawydoff (*1909–52*) on regeneration and embryology, Stiasny-Wijnhoff (*1909–42*) on taxonomy, anatomy and zoogeography, Brinkmann (*1912–32*) on the pelagic hoplonemerteans, and Schmidt (*1923–62*) on reproduction and embryology.

In more recent years the notable works of such authors as Corrêa, Friedrich, Gontcharoff and Iwata have become well established;

[1] Dates listed in italics refer to the period during which the author published on nemerteans.

during the last decade valuable publications by Bierne, G. J. Müller and Kirsteuer have added to the growing volume of current literature on nemerteans, as well as those of many other authors too numerous to cite here. Friedrich (1965, 1969) has produced two very useful papers that will be of benefit to those interested in nemertean research. In these he lists and classifies the majority of papers and books relating to nemerteans; the reader is referred to Friedrich's work for a far more detailed bibliography than can be contained in a book of this nature.

During the last few years such developments in zoological research as the elaboration of more sophisticated histochemical and physiological techniques, and the advances in electron microscopy, have opened up whole areas of investigation hitherto unexplored. That workers are employing these approaches in nemertean studies is a promising sign – nemerteans are rewarding and satisfying animals to work with, and there is still so very much that we do not know about them. For example, how many texts on general littoral ecology even mention them, let alone deal with them at length?

CLASSIFICATION AND MORPHOLOGY

Nemertean worms may be defined as unsegmented, bilaterally symmetrical, acoelomate animals, with a gut possessing separate mouth and anus, a blood vascular system, and a characteristic eversible proboscis situated dorsal to the gut in an enclosed tubular cavity, the rhynchocoel.

The phylum can be conveniently split on morphological grounds into two classes, each of which is further subdivisible. There is some disagreement as to whether or not the two principal divisions warrant class status in view of the great similarity in structure found throughout the phylum. Coe (1943), for example, regards the two groups as distinct classes, whereas Hyman (1951) feels that their members are not sufficiently dissimilar to rate a higher division than into subclasses. Hyman's reasoning would be acceptable if nemerteans were regarded as a class of the phylum Platyhelminthes, to which they are undoubtedly related, but since it is now generally recognised that they are sufficiently advanced to be separated from the platyhelminths, the classification employed in this book is based upon that given by Coe.

PHYLUM: RHYNCHOCOELA (NEMERTEA)

Class I: ANOPLA

Mouth below or posterior to the cerebral ganglia; central nervous system situated within the body wall (epidermis, dermis, or body-wall musculature); proboscis not differentiated into three regions and either not armed or provided with large numbers of rhabdite-like epithelial barbs.

Order 1: PALAEONEMERTEA (PALAEONEMERTINI) – body-wall musculature either of two layers (outer circular and inner longitudinal) or three layers (outer circular, middle longitudinal, inner circular); dermis gelatinous or absent; central nervous system either in inner longitudinal musculature or external to body-wall muscles.

Order 2: HETERONEMERTEA (HETERONEMERTINI) – body-wall musculature of three layers (outer longitudinal, middle circular, inner longitudinal), sometimes with additional thin inner circular and outer oblique muscle layers; dermis well developed and fibrous; central nervous system in middle circular muscle layer.

Class II: ENOPLA

Mouth anterior to cerebral ganglia; central nervous system internal to the body-wall musculature, which is two-layered (outer circular, inner longitudinal); proboscis regionally differentiated and armed with one or more needle-like central stylets (except in the order Bdellonemertea).

Order 3: HOPLONEMERTEA (HOPLONEMERTINI) – proboscis armed with one or more central stylets; intestine straight with paired lateral diverticula; no posterior ventral sucker.
    Suborder 1: MONOSTYLIFERA – central stylet armature a single structure carried in a large cylindrical basis.
    Suborder 2: POLYSTYLIFERA – proboscis armature consists of a pad or shield bearing numerous small stylets.
        Tribe 1: REPTANTIA – body adapted for crawling or burrowing; rhynchocoel with caecal outgrowths; cerebral organs and nephridial system present.
        Tribe 2: PELAGICA – bathypelagic, with the body adapted for free swimming or floating in deep water; rhynchocoel without caeca; cerebral organs and nephridial system absent.

Order 4: BDELLONEMERTEA (BDELLONEMERTINI or BDELLOMORPHA) – proboscis not armed with stylets; intestine sinuous and without lateral diverticula; foregut doliiform, with many papillae; with a posterior ventral sucker.

A more recent classification, proposed by Iwata (1960a), is based not only on morphological but also on embryological relationships. Iwata's divisions, all falling below class level, are:

ANOPLA: Order 1: ARCHINEMERTEA

Order 2: PALAEONEMERTEA
Order 3: HETERONEMERTEA
ENOPLA: Order 4: HOPLONEMERTEA
Suborder 1: MONOSTYLIFEROIDEA
Suborder 2: POLYSTYLIFEROIDEA
Suborder 3: BDELLONEMERTOIDEA

The differences between Iwata's and Coe's classifications are in the establishment of a new anoplan order, the Archinemertea, and in the relegation of the bdellonemerteans to a subordinate level. The order Archinemertea is erected for the single family Cephalothricidae, previously included in the Palaeonemertea. Iwata defines members of the order as: having a sharply pointed head, with the mouth situated far behind the brain; body musculature with two layers (outer circular, inner longitudinal); central nervous system lying in the inner longitudinal musculature; cerebral organs and dorsal blood vessel absent; excretory system consisting of many isolated nephridia.

The elimination of the order Bdellonemertea is based upon Bürger's (1895) comment that the group merely represents a family of the Hoplonemertea. Iwata points out that these two groups possess the same embryology and have their central nervous system in an identical position, namely, internal to the body-wall muscles, surrounded by parenchyma. Morphological differences shown by bdellonemerteans he believes are secondarily derived from the group's adoption of parasitic (commensal) habits.

The basis of Iwata's proposition appears to rest upon the prime degree of importance placed on the position of the nervous system compared to other morphological features. The absence of cerebral organs, for example, used as one criterion for separating the Cephalothricidae from the remaining Palaeonemertea, becomes significantly less important when it is appreciated that certain hoplonemerteans too lack these structures. In addition, other palaeonemertean examples, such as *Carinoma*, *Carinina* and *Tubulanus* (= *Carinella*), have their organs confined to simple sensory pits, situated upon the margins of the head, which are quite different from the more typical structures found in other species.

The bdellonemerteans are morphologically far more different from any other nemertean group than are the Cephalothricidae from the remaining palaeonemerteans, and it seems somewhat incongruous to, on the one hand, relegate, and on the other, upgrade, the respective groups under these circumstances. In the absence of sufficient justification, Iwata's proposed classification is not adopted in this book.

A list of the currently recognised nemertean genera is provided in the Appendix.

External features

Nemerteans are for the most part elongate, vermiform animals with a soft body capable of extreme contraction and elongation. Some species, especially amongst the palaeonemerteans, are exceedingly long and slender (*Cephalothrix, Paralineus, Procephalothrix, Tubulanus*) (Fig. 1A), but most are less attenuated and relatively broader with a body that is somewhat dorsoventrally compressed and rather ribbon-like (*Cerebratulus, Euborlasia*) (Fig. 1B), wide and flattened (*Dinonemertes, Drepanophorus, Nectonemertes*) (Fig. 1C), leech-like (*Malacobdella*) (Fig. 1D) or, much more commonly, with a long or short body that in section is cylindrical (*Lineus, Obürgeria, Oerstedia, Nemertopsis, Tetrastemma*) (Fig. 1E), elliptical (*Amphiporus, Dushia, Flaminga, Hubrechtella*) (Fig. 1F), or convex above and flattened below (*Emplectonema, Uchidana*).

Great variation is found in the sizes of nemerteans, their length ranging from only a few millimetres (*Carcinonemertes, Coenemertes, Prostoma, Tetrastemma*) up to several metres (*Cephalothrix, Lineus, Tubulanus*). Most species are less than about 20 cm long, and few exceed half a metre. The largest specimen yet reported is the heteronemertean *Lineus longissimus*, examples of which regularly attain lengths of several metres, found on the east coast of Scotland. A conservative estimate of one specimen indicated a body some 30 metres long. Despite the range of lengths found in the phylum, the vermiform species rarely exceed a few millimetres width. The larger and broader pelagic hoplonemerteans are regularly 1 to 2 cm wide, a size also attained by the heteronemertean *Cerebratulus marginatus*, but none of these approaches the 45 mm and 56 mm breadths of *Euborlasia maxima* and *Dinonemertes investigatoris* respectively, the two bulkiest nemertean species yet found (Coe, 1905a; Brinkmann, 1917).

The determination of nemertean sizes is, in fact, problematical in view of their great powers of extension and contraction, and it is virtually impossible to establish a norm or average length. Ideally, living specimens should be measured during locomotion when partially extended, or else should have maximal and minimal dimensions recorded, whereas preserved animals must, in general, be presumed to be at least partially contracted. Because of these difficulties, relatively little emphasis can be placed upon body dimensions when identifying nemerteans.

Anteriorly nemerteans are pointed, rounded or blunted, a distinct

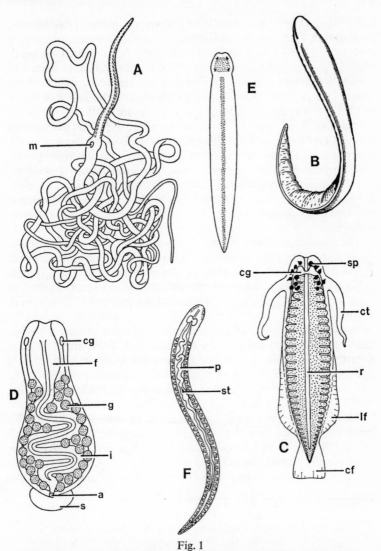

Fig. 1

The body shape of nemerteans. **A**, *Cephalothrix major*; **B**, *Cerebratulus latus*; **C**, *Nectonemertes mirabilis*; **D**, *Malacobdella grossa*; **E**, *Tetrastemma piolinum*; **F**, *Amphiporus lactifloreus*. **a**, anus; **cf**, caudal fin; **cg**, cerebral ganglia; **ct**, cephalic tentacle; **f**, foregut; **g**, gonad; **i**, intestine; **lf**, lateral fin; **m**, mouth; **p**, proboscis; **r**, rhynchocoel; **s**, sucker; **sp**, spermary; **st**, stylet apparatus. (A redrawn from Coe, 1930b; B redrawn from Coe, 1905a; C after Coe, 1926; E redrawn from Corrêa, 1957)

head being absent. Representatives of several genera (*Amphiporus,*
Itanemertes, Tetrastemma, Tubulanus, Zygeupolia) possess an anterior
cephalic lobe that is lanceolate, spatulate, heart-shaped or semi-
circular in shape (Figs. 2A–E), but this does not constitute a head as
it most frequently does not incorporate the cerebral ganglia. Many
nemerteans possess at their anterior ends shallow transverse or longitudinal grooves, the cephalic grooves. These are most striking in the
heteronemerteans, especially the lineids, where they form deep lateral
furrows known as cephalic slits (Figs. 2F,N). Eyes are present at the
anterior end, the number varying both inter- and intraspecifically,
and ranging from two (*Amphiporus, Carcinonemertes, Poseidone-*
mertes), four (*Algonemertes, Lineus, Tetrastemma*), six (*Cerebra-*
tulus, Prostoma), many (*Emplectonema, Proneurotes*), to over one
hundred (*Zygonemertes*) (Figs. 2A,E,G–K). Large specimens of *Amphi-*
porus formidabilis may possess more than 250 eyes.

The eyes are arranged more or less bilaterally and, with the exception of the genus *Zygonemertes*, do not usually occur behind the
cerebral ganglia. When the number of eyes is small (2–4), it is usually
constant for a given species. A few nemerteans are totally eyeless, at
least as adults (*Cephalothrix, Malacobdella, Potamonemertes*) (Figs.
1A,D).

In some of the bathypelagic genera a single pair of lateral cirri or
tentacles (called nuchal cirri by Coe and Ball, 1920) are positioned
near the anterior end of the body. These are found in both sexes of
Balaenanemertes, but only in males of *Nectonemertes* (Fig. 1C), and
are clearly secondary sex characters (Foshay, 1912; Brinkmann,
1917; Coe, 1936, 1945a).

The proboscis pore usually opens at or just ventral to the anterior
tip, but may open from the surface of the cephalic lobe when one is
present. The rounded or slit-like mouth is positioned ventrally, either
immediately behind the cerebral ganglia in the Anopla (Fig. 6B) –
except for the palaeonemertean family Cephalothricidae, where it is

Fig. 2
Anterior anatomy. **A,** *Itanemertes nonatoi;* **B,** *Tetrastemma nigrifrons;*
C, *Tubulanus annulatus;* **D,** *Carinella (Tubulanus) rubra;* **E,** *Amphiporus*
bimaculatus; **F,** *Lineus ruber;* **G,** *Carcinonemertes carcinophila;* **H,**
Prostoma jenningsi; **I,** *Emplectonema bürgeri;* **J,** *Zygonemertes vires-*
cens; **K,** *Amphiporus formidabilis;* **L,** *Uchidana parasita;* **M,** *Dushia*
atra; **N,** *Corsoua kristenseni.* **cg,** cerebral ganglia; **cgr,** cephalic groove;
cs, cephalic slit; **e,** eye; **ln,** lateral nerve cord; **m,** mouth; **pp,** proboscis
pore. (A redrawn from Corrêa, 1957; B, D, E, I, J, K, redrawn from
Coe, 1905a; G modified from Humes, 1942; L redrawn from Iwata,
1967; M after Corrêa, 1963; N redrawn from Corrêa, 1963)

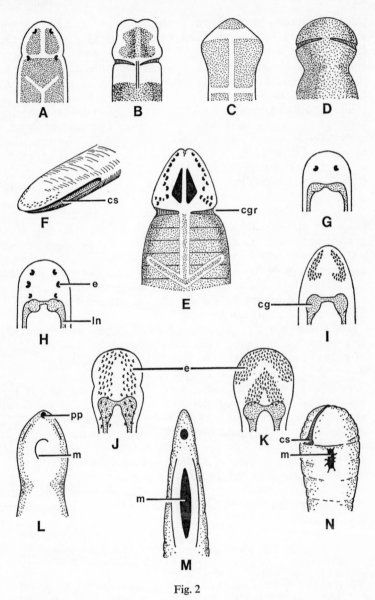

Fig. 2

placed much more posteriorly at the end of an extended and narrow snout (Figs. 1A, 6A) – or in front of the cerebral ganglia in the Enopla, either opening immediately behind the proboscis pore, or else uniting internally with the proboscis apparatus and opening to the exterior via a common rhynchodaeal aperture (Fig. 6C). In bdellonemerteans the rhynchodaeum is lost during development, the proboscis opening into the dorsal surface of the foregut and sharing with the gut a common anterior aperture. The unusual hetero-nemertean *Uchidana parasita* (Iwata, 1967) has an asymmetrical bow-shaped mouth with thick left and slender right lips (Fig. 2L). An extremely large mouth is found in the heteronemertean *Dushia atra* (Corrêa, 1963), where it consists of a very long, thick-lipped ventral slit (Fig. 2M).

The sides of the body are mostly more or less smooth in outline, but a few species, particularly those that are dorsoventrally com-pressed and rather ribbon-like (*Cerebratulus, Parapolia*), have their lateral margins folded and quite irregular in shape. In paler coloured nemerteans, or in those where the ripe gonads possess a different colouration from the remainder of the body, the internal pseudo-metamerism exhibited by the gonads and intestinal diverticula is often discernible through the body wall.

The posterior end of nemerteans normally either tapers gradually to a sharp or blunt tip or, as in many heteronemerteans, terminates in a small tail or caudal cirrus (caudicle, Coe, 1905a) a few milli-metres long (*Cerebratulus, Flaminga, Micrura, Zygeupolia*) (Figs. 3E,H, 4E). In swimming forms, most strikingly in the bathypelagic hoplonemerteans, the posterior end is considerably flattened and extended into a simple or bilobed caudal fin, as in *Balaenanemertes, Nectonemertes* and *Proarmaueria* (Figs. 3A,B). This may, in turn, be anteriorly developed into thin lateral margins extending a greater part of the body length, as in *Nectonemertes* (Coe and Ball, 1920). The anus opens at or just dorsal to the posterior end, or at the base of the caudal cirrus, usually dorsally, when one is present. In the bdellonemerteans (*Malacobdella*) there is a single posterior ventral

---

Fig. 3
Body shape and colour patterns. **A**, *Balaenanemertes chavesi*; **B**, *Proarmaueria pellucida*; **C**, *Amphiporus angulatus*; **D**, *Tetrastemma signifer*; **E**, *Micrura verrilli*; **F**, *Tetrastemma reticulatum*; **G**, *Micrura pardalis*; **H**, *Tubulanus rhabdotus*; **I**, *Evelineus tigrillus*. **cc**, caudal cirrus; **cf**, caudal fin; **cg**, cerebral ganglia; **cs**, cephalic slit; **ct**, cephalic tentacle; **i**, intestine; **p**, proboscis; **r**, rhynchocoel; **sp**, spermary. (**A** redrawn from Coe, 1945a; **B** redrawn from Coe, 1926; **C–G** redrawn from Coe, 1905a; **H** redrawn from Corrêa, 1954; **I** after Corrêa, 1954)

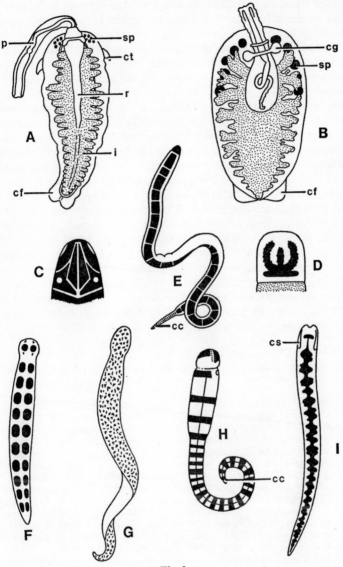

Fig. 3

sucker (Figs. 1D, 6D), used for attachment to the mantle lining of the bivalve hosts. In these forms the anus opens at the posterior end of the body but dorsal to the sucker.

The majority of nemerteans are more or less uniformly coloured in various shades of red, brown, green, orange, pink, yellow or grey, but several species are strikingly and characteristically marked by distinctive, and often specific, patterns (Plate 1, Figs. 3C–I). These may comprise stripes and bands, or both, contrasting colours arranged into geometric shapes, especially anteriorly and dorsally, or speckles or a marbled effect appearing over most of the body surface. Commonly the dorsal surface is either more deeply pigmented or differently coloured than the ventral, and patterns, when present, tend to be restricted to the upper surface or more clearly marked there. Many of the bathypelagic species are brilliantly coloured in red, orange, scarlet or yellow. A few nemertean species are white or creamish in colour, or else translucent with the body organs clearly visible through the epidermis. The colouration often depends upon the sex of the animal concerned, particularly in the breeding season when changes in colour may be attributed to pigments contained within the ripe gonads. In some species (*Amphiporus, Lineus*) the region of the cerebral ganglia is marked by a pinkish tinge, but in others this may be partially or totally obscured by the general body pigmentation.

The body wall

The body wall of nemerteans comprises three distinct components,

---

Fig. 4

Internal anatomy, epidermis, muscle systems and caudal cirrus. **A**, Cellular elements of the epidermis; **B**, A ciliated epidermal cell; **C**, Composite gland of *Amphiporus gelatinosus*; **D**, Rhabdite cells of *Prostoma jenningsi*; **E**, Caudal cirrus of *Zygeupolia litoralis*; **F**, Dorsoventral muscles in *Neuronemertes aurantiaca*; **G**, Body-wall structure of *Amphiporus lactifloreus*. a, anus; bc, basal cell for replacement of gland cells; bk, basal knob of cilium; cbl, cirrus blood lacuna; clc, ciliated cell; cm, circular musculature; cmf, circular muscle fibre; cn, cirrus nerve; cnc, cytoplasmic nuclear connection; dn, dorsal nerve; dvm, dorsoventral muscles; ebm, epidermal basement membrane; ep, epidermis; er, epidermal rhabdite; gc1, gland cell containing coarse granular secretion; gc2, gland cell containing homogeneous secretion; i, intestine; lbv, lateral blood vessel; lm, longitudinal musculature; lmf, longitudinal muscle fibre; ln, lateral nerve cord; mbv, middorsal blood vessel; my, myoseptum; n, nucleus; raf, radial argentophil fibres; rm, radial membrane; sec, sensory epidermal cell; smg, submuscular composite gland; ubc, undifferentiated basal cell. (A redrawn from Coe, 1905a; B, E modified from Thompson, 1901; C after Coe, 1905a; F redrawn from Coe, 1926; G redrawn from Cowey, 1952)

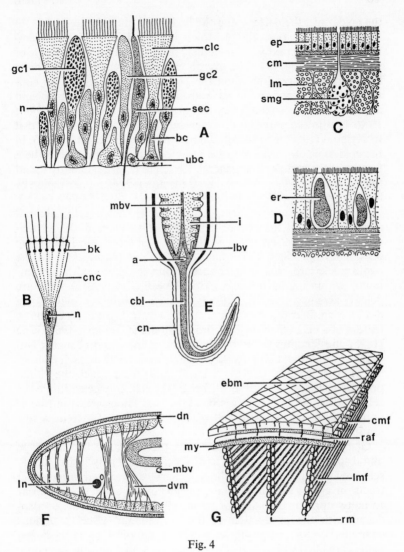

Fig. 4

the epidermis, the dermis, and the body-wall musculature.

The epidermis, one cell-layer thick, is formed from ciliated columnar cells interspersed with glandular, sensory and interstitial cells of several types (Fig. 4A). The overall epidermal height may be quite low, variable depending upon the body region concerned, or considerably thickened with its cells correspondingly narrow and attenuated. Humes (1942) records an epidermal thickness of about 10 to 11 μ in the parasitic *Carcinonemertes carcinophila*, in contrast to Hylbom (1957) who found a maximum cellular height of 171 μ in the epidermis of *Carinina coei*, and noted that this appeared to be a feature of the carininids in general. Kirsteuer (1967a) similarly found that in *Hubrechtella sarodravayensis* the anterior glandular region of the epidermis is particularly thick.

The major epidermal components, the epithelial cells, are tall ciliated structures, usually with broad distal ends and slender filamentous bases that are often twisted amongst those of neighbouring cells. They are filled with a fine to coarse cytoplasm in which the ovoid nuclei are usually positioned within the proximal half, commonly within the narrowed neck of the cell (Coe, 1905a). Cilia are densely arranged over the entire body surface, being normally some 8–12 μ long. Between the cilia, microvilli (Plate 3A), a fairly common feature of a ciliated epithelium, are distributed (Jennings, 1969). Coe (1895) and Thompson (1901) describe the epidermal cilia of *Cerebratulus lacteus* and *Zygeupolia litoralis* respectively as each consisting of several parts, namely, a basal knob from which the cilium projects resting on the distal surface of the cell, connected inside the cell by a slender thread to a second, smaller knob, which in turn is apparently linked with the cell nucleus by a line of extremely fine granules (Fig. 4B).

Gland cells are distributed between the ciliated cells, their type and disposition varying in different species. They may be irregularly scattered over most or all of the body surface, arranged into distinct tracts, or aggregated into definite regions of the body. There appear to be two principal epidermal gland cell types in nemerteans, associated with the production of the mucoid and viscid secretions characteristic of the group as a whole. Elliptical or flask-shaped glands are filled with a coarsely granular or foamy mucous secretion, whereas the elongate, rod-like glands contain a much more homogeneous serous substance. A third gland type, the so-called clustered or packet gland cells of Bürger (1895) are positioned more deeply within the epidermis but open to the surface via a common duct, as in some species of *Tubulanus* (Coe, 1905a). Hylbom (1957), in stating that these glands are unicellular, agrees with Bergendal (1902) and

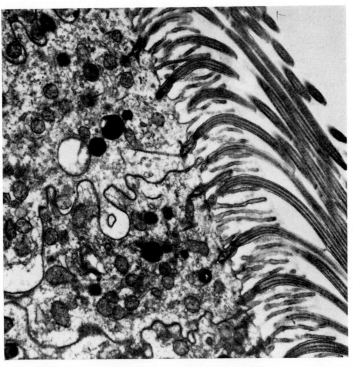

Above: Plate 3A
Electronmicrograph of
distal epidermal region
of *Lineus ruber*, showing
cilia, a ciliary rootlet,
microvilli and mito-
chondria. Photograph by
Dr J. B. Jennings.

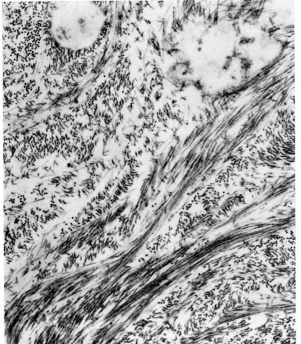

Left: Plate 3B
Electronmicrograph of
dermis of *Amphiporus
pulcher*, showing the
arrangement of filaments.
Photograph by Dr K. J.
Pedersen (from Pedersen,
1968).

Plate 4A
Electronmicrograph of
filamentous connective tissue
membranes (arrowed)
enclosing various organs of
Amphiporus pulcher. GO.,
cells of gonad; GS., ground
substance; NE., nerve cells.
Photograph by Dr K. J.
Pedersen (from Pedersen,
1968).

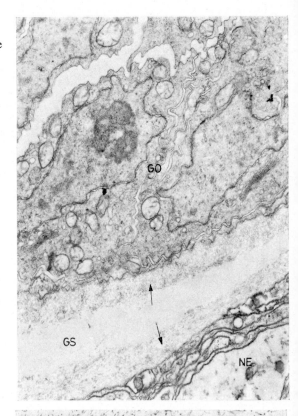

Plate 4B
Electronmicrograph of
closely apposed muscle cells
of *Lineus bilineatus*. Cells are
situated in the composite
dermis and are surrounded
by a ground substance
containing filaments.
Hemidesmosomes indicated
by arrows. Photograph by
Dr K. J. Pedersen (from
Pedersen, 1968).

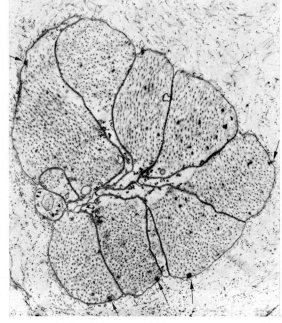

Nawitzki (1931) so far as the genus *Carinina* is concerned, but Coe (1943) notes that, in some palaeonemerteans at least, the glands are in fact composite, consisting of several cells sharing a single secretory duct (Fig. 4c). In other nemertean groups the packet glands, whilst retaining their opening to the body surface, may be placed in the dermis, amongst the muscle layers, or submuscularly in the parenchyma (Coe, 1905a; Alvarado, 1956; Brunberg, 1964). They are then known as subepidermal, subepithelial, submuscular or cutis glands. Coe (1905a) observes that in many hoplonemerteans the submuscular glands are more or less numerous, situated in the parenchyma between or beneath the muscle layers of the body wall. They are usually restricted to the cephalic region of the body, but may extend posteriorly to the end of the oesophageal region (*Amphiporus nebulosus*) or even as far back as the intestine (*Emplectonema bürgeri*).

Some hoplonemertean species (*Amphiporus, Emplectonema, Prostoma, Zygonemertes*) have rod-like or sickle-shaped, apparently glandular, structures distributed throughout the epidermis (Fig. 4D). These appear to be the same as the gland cells containing rod-like bodies found in *Carinina* species (Hylbom, 1957) that Bergendal (1902) compared with platyhelminth rhabdites. The function of these calcium-containing bodies in nemerteans has not yet been determined, but in some species they are so abundant that their slightly yellowish colouration affects the overall body hue.

The interstitial cells are small, branched and often anastomosing structures, or else form a basal syncytium to the epidermis that extends between the narrow proximal regions of other cells present. In many nemerteans it is these cells that contain the pigment granules responsible for body colouration and patterns, but in others the pigment granules are found much more deeply, either in the dermis or amongst the body-wall muscle layers. Coe (1905a) states that some of the interstitial cells seem to correspond to elements of the connective tissue, whilst others are strictly epithelial.

Specialised slender sensory cells with a single flagellum are always present in the epidermis (Fig. 4A), often being aggregated into special sense organs restricted mainly to the anterior regions of the body. Rather longer cilia (or setae), presumably tactile in function, are found in certain species of *Prostoma* (Böhmig, 1898) and *Ototyphlonemertes* (Corrêa, 1954), confined to very precise parts of the body (Fig. 12E). These structures are extremely difficult to discern even in living material and may well be of more general distribution in the phylum.

Beneath the epidermis there occurs a layer of connective-like tissue that forms the dermis or cutis, defined by Bürger (1895) as the sub-epithelial glandular layer, usually containing numerous muscle fibres, that divides the epidermis from the body-wall musculature. This layer is very thick in many bathypelagic hoplonemerteans, being externally developed into numerous hexagonal, pentagonal or irregular cup-shaped pits or convolutions for the attachment of the thin and deli-cate epidermis (Coe, 1926).

Two principal types of this tissue are generally recognised in nemerteans, one a thick to thin layer of homogeneous hyaline connective tissue composed of a gelatinous ground substance in which nuclei and a variety of delicate fibres and cells are embedded, the other a thick zone of predominantly fibrous connective tissue (Plate 3B). The former is found in all orders except the Heter-onemertea, the latter being characteristically associated with this group. Coe (1943) differentiates between these two types by calling the former a basement layer, the latter a cutis, but Hyman (1951) correctly points out that since both are in fact forms of connective tissue, an observation confirmed by Pedersen (1968), and since sub-epidermal connective tissue throughout the animal kingdom is known as dermis, she considers it only logical to apply the term to both types of nemertean subepidermal tissue rather than to adopt the divisions of Coe.

The fibrous dermis often contains bodies of both simple and packet glands, as well as muscle fibres. In *Cerebratulus* and many other lineids, as well as in the Baseodiscidae, the region is differentiated into two zones. Of these the outermost is glandular, the innermost muscular with oblique, longitudinal and circular fibres. Certain glands, especially of the composite type, may penetrate even deeper into the body wall and extend into the general parenchyma internal to the main musculature. Such is the case in *Baseodiscus univittatus*. Some heteronemerteans have the development of longitudinal muscles so numerous in the innermost dermal zone that the limit between this and the outer longitudinal musculature of the principal muscle layers is somewhat obscured.

Gontcharoff and Lechenault (1966) recognise in the heteronemer-teans *Lineus ruber* and *L. viridis* six types of subepidermal gland cells on the basis of their histochemical differences. Of these cells two types are mucus secreting (acid and neutral mucus), the remaining four being grouped together as serous glands but divisible on histo-logical as well as histochemical grounds. The histochemical charac-teristics of the serous glands are summarised in Table 1.

Ultrastructural investigations on the formation of these gland types reveal that the acid mucous glands, particularly evident in the

Table 1

The histochemical characteristics of lineid serous glands (from Gontcharoff and Lechenault, 1966).

	Cell type			
Radical	With homogeneous cytoplasm	With granular cytoplasm	Rhabdites	From gravid females
Amino	+ +	+ +	+	+
Sulphydryl	+ +	+	+ +	±
Phenol	+ +	+	+	+
Guanidyl	+ +	+	±	−
Indolyl	+ +	+	±	−
Bromphenol blue	+	+	+ +	±

+ + intense reaction   + definite reaction   ± weak reaction   − negative

precerebral region of the body, contain a characteristic single or double Golgi apparatus and a secretion that is reticulate in appearance. In addition the histochemical differentiation of the serous glands is confirmed by differences in their ultrastructural appearance.

Internal to the dermis is a zone of thick and powerful musculature, the body-wall muscles. Its precise layer arrangement forms one of the diagnostic features in the classification of nemerteans (Fig. 5). Palaeo-, hoplo- and bdellonemerteans have two layers, consisting of outer circular and inner longitudinal fibres, although two additional zones of diagonal or spiral fibres are placed between the two principal layers in many species of Palaeo- and Hoplonemertea. Spirally orientated fibres run in opposing directions in the two zones. In species of *Cephalothrix* and in the heteronemertean *Lineus socialis*, which are able to coil their bodies into particularly tight spirals, these muscle regions are especially well developed. In addition some palaeonemerteans (*Carinoma*, *Tubulanus*), as well as some heteronemerteans, have an extra layer of circular fibres running internal to the longitudinal zone. This, when present, may extend for the full body length or may be confined to particular, usually anterior, regions associated with the foregut. The inner and outer layers of circular fibres may be transversely joined by fibrous cross connections penetrating the longitudinal zone (Coe, 1943).

Heteronemerteans possess a third distinct layer of longitudinal fibres external to the other two, in which case the spiral or diagonal fibres, when present, are found between the outer longitudinal and middle circular layers.

The body-wall muscle layers are more or less uniform in the

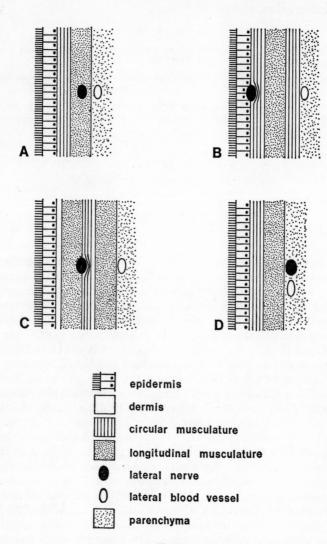

Fig. 5
Systematic relationships of the body-wall muscle layers, lateral nerve cords and lateral blood vessels. A, B, Palaeonemertea (A is equivalent to the Archinemertea of Iwata, 1960a); C, Heteronemertea; D, Hoplonemertea and Bdellonemertea

anterior portion of the body, but in some forms the longitudinal muscles show a stronger development and serve as retractors of the head end, the circular musculature being correspondingly reduced. In heteronemerteans much of the anterior region is filled with a dense meshwork of fibres of which the major type is longitudinal, a comparable situation also occurring in some hoplonemerteans but with the fibres arranged more definitely into a precerebral transverse septum. The body-wall musculature throughout the body is irregularly penetrated by radial fibres that extend to or into the epidermis. These are called connective intermuscular fibrils by Alvarado (1956).

Some modification of the muscular arrangement is found in a number of species. In those heteronemerteans with an extra inner circular layer (species of *Micrura* and *Zygeupolia*), this musculature is entirely limited to the extreme posterior end of the foregut region, where it serves as a sphincter (Coe, 1943). A longitudinal muscle plate, derived from the inner longitudinal zone of the body-wall musculature, is present in *Hinumanemertes*, where it extends posteriorly between the proboscis and the gut (Iwata, 1970). The caudal cirrus of *Micrella* (Punnett, 1901a) and *Zygeupolia* (Thompson, 1901) has both circular and longitudinal muscles, more or less continuous with the two inner zones of the main body-wall musculature. In addition it contains slender prolongations of the lateral nerve cords, and a large central blood lacuna joined to both the lateral and middorsal blood vessels (Fig. 4E).

All nemerteans have dorsoventrally orientated muscle fibres extending through the parenchyma between the dorsal and ventral portions of the body-wall musculature. These muscles frequently alternate with the intestinal diverticula and, as such, reflect a pseudometameric arrangement. They are best developed in species with broad flattened bodies, especially in the swimming and bathypelagic nemerteans (Brinkmann, 1917; Coe, 1926, 1927a, 1936) (Fig. 4F). In other, benthic, examples that are capable of swimming (*Cerebratulus lacteus*), the body is dorsoventrally compressed and double bands of muscles occur between the intestinal diverticula, forming two sheets of muscular tissue that straddle the gonads (Coe, 1895). In general the more the species is adapted for swimming, the stronger is the development of these dorsoventral muscles, their greatest size being found in the caudal fins of bathypelagic species such as *Nectonemertes mirabilis* and *N. pelagica* (Cravens and Heath, 1906; Coe and Ball, 1920).

Several pelagic hoplonemerteans are adapted more for passively floating than for actively swimming (*Mergonemertes, Pelagonemertes*,

B

Planktonemertes, *Planonemertes*). In these forms there is a corresponding reduction in the body-wall musculature, particularly of the circular zones which may be present only as isolated bands of muscles, separated by parenchyma (Coe, 1926). Often the musculature is restricted to the development of dorsal and ventral plates of longitudinal fibres. In fact, the body-wall layers of the bathypelagic forms are relatively thin compared with those of the littoral species.

In the floating Pelagica the great development of the gelatinous parenchyma seems to enhance their floating ability by reducing the bodies' specific gravity (Coe, 1927a). In some forms (Armaueriidae, Pelagonemertidae, *Pendonemertes*) a narrow band of muscle fibres splits off on each side of the body from the layers lining the proboscis sheath, immediately behind the cerebral ganglia. These extend posteriorly for the remaining body length in close connection with the lateral nerve cords. They are mostly very thin, only being one or two fibres thick, but in a few, such as *Balaenanemertes chuni*, are so thick as to be equal in diameter to the lateral nerves (Coe, 1926).

In nemerteans each muscle fibre consists of a single slender, and often filiform, contractile cell that is provided with an extremely small and slender nucleus, surrounded by a minute amount of undifferentiated cytoplasm (Coe, 1905a). The muscles are of the smooth type.

Cowey (1952) investigated in detail the structure and function of the basement layer (dermis) and body-wall musculature as a complete system in the hoplonemertean *Amphiporus lactifloreus*. He found that it consisted of several distinct parts (Fig. 4G). The outermost region, situated dermally, comprises a five- to six-layered zone of alternate right- and left-hand geodesic fibrous helices arranged into a lattice, immediately beneath the epidermis. The sides of the lattice parallelograms measure 5–6 μ in length, with the fibres being attached to each other where they cross. By this means fibres are made incapable of slipping relative to each other. Inside this region a zone of circular muscle fibres containing two systems of argyrophil fibrils (i.e., fibrils having a special affinity for silver) runs around the body. In one system the argyrophils run parallel to the muscle fibres at intervals of some 10 μ, the other has its fibrils extending through the muscle zone from the dermis to the myoseptum that divides the two major muscle blocks of the body. The myoseptum possesses a structure identical with that of the epidermal basement lining.

The innermost, longitudinal, muscle layer contains fibres arranged in strata on either side of a series of longitudinal radial membranes.

These findings are confirmed by Alvarado (1956) and Clark and Cowey (1958).

Connective tissue

Several sites for connective tissue can be found in nemerteans, including those present in the dermis, situated between muscle layers, forming the general membranes of the body that surround the blood vessels, nerves, nephridia and other structures, or serving as the general packing tissue filling the space between gut and body wall. The most recent study of nemertean connective tissues (Pedersen, 1968) suggests that since the system in all probability forms a unity in the animals, no classification into types need really be attempted. The entire connective tissue region of nemerteans corresponds to the parenchyma or mesenchyme of flatworms, but exhibits a far greater complexity. It thus both shows a continuity with the turbellarian level of organisation at the same time as it includes elements conforming with patterns found in most animal groups, including vertebrates.

Pedersen recognises three connective tissue elements (Plate 4A). These are a ground substance, filaments and several types of free cells. The amount of ground substance present varies according to the species concerned; it may be fairly abundant, as in *Amphiporus pulcher*, or so scanty that the body-wall musculature lies virtually in contact with the gut epithelium. It consists of a homogeneous to finely fibrillar material in which occur the various cellular and fibrous connective tissue components. The inclusions are mostly sparsely distributed, but there is a tendency, in some species at least, for single muscle cells to be particularly abundant in the head region. The ground substance is apparently composed of both acid and neutral mucopolysaccharides, these probably occurring in low concentrations. A distinctive proteinaceous component is also present.

The filaments occurring in connective tissue are essentially the same as those described by Cowey (1952), although Pedersen has extended the information on these structures to include those present in the general body parenchyma. Dermal filaments often run together in bundles to form fibrils of somewhat variable diameter, commonly in the range $0\cdot3-0\cdot6\,\mu$. They are not in any way regularly arranged within the dermis, some being orientated longitudinally, others circularly, and yet others quite irregularly. Some fibril bundles bend and change their plane of orientation. Quite often filaments are found in close association with the proximal regions of the epidermis, but do not apparently penetrate it. In the body, filaments often form a continuity with those of the dermis, but tend to be much more highly developed and ramify throughout the animal. The main fibrillar system accounts for a fairly high proportion of the body mass and is, when differentiated into distinctive membranes or sheaths, responsible for the tissue investing the various body organs and structures. These membranes may be several μ thick. Individual muscle cells are only rarely ensheathed, the cell membranes of the muscle cells either adjoining directly the ground substance or being closely apposed to neighbouring muscle cells.

In the heteronemertean *Lineus bilineatus*, which has an obviously composite dermis with large areas of ground substance disposed between the longitudinal muscle cells, up to a dozen of these cells may run together in close apposition. A highly characteristic feature of these bundles is the presence of hemidesmosomes (desmosomes or plasmodesmata are bridges of cellular material joining adjacent cells; at the external periphery of the muscle bundles a 'half-bridge' or hemidesmosome is present) at the junction between ground substance and muscle-cell membranes (Plate 4B). There is no attachment between these hemidesmosomes and connective tissue filaments, and desmosomes are never seen occurring between the muscle cells themselves. Only extremely rarely have comparable microstructures been seen in other species.

Perhaps the most important feature of the connective tissue filament system is that it forms a continuous fibrillar arrangement penetrating the body of the worms and binding together the various organ systems, forming part of a 'collagen skeleton' that Pedersen has also found in triclad and polyclad turbellarians.

Pedersen tentatively classes the cellular components of the connective tissue into three main groups. The principal characteristics of the groups are shown in Table 2.

The group 3 cells are far more heterogeneous than those of the other two groups, and in fact a considerable overlap in form occurs between components of groups 1 and 3.

Connective tissue cells are extremely variable in nature, and no absolute division between the various types can be achieved. Often several functions are combined within a single cell, for example, the synthesis of extracellular substances, the phagocytosis of decaying cellular products, and the carriage of nutritive material for use in the various regenerative mechanisms. During regeneration or starvation many of the connective tissue cells become heavily loaded with pigment granules; these correspond to the so-called 'wandering cells' described by Nusbaum and Oxner (1910).

In the majority of nemerteans the anterior parenchyma contains clusters of gland cells, the cyanophilous glands of Hyman (1951). These may extend some distance posteriorly, or be restricted to the extreme anterior tip. These are the frontal or cephalic glands, described in more detail under the section on sense organs. A few

Table 2
Characteristics of the connective tissue cellular components (after Pedersen, 1968).

	Cell group		
	1	2	3
Site in body	Mainly dermis, but also in areas with ground substance and in strands of fibrils, e.g., those surrounding the gastrodermis.	Particularly evident in longitudinal muscle layer.	Distributed in areas with abundant ground substance and between muscle cells, especially in longitudinal layer. Sometimes closely apposed into groups.
Nucleus	Rather small, outline irregular, elongated or round with much evenly distributed chromatin. Sometimes with nucleoli.	Very variable in size and appearance, round, elongate, irregular or triangular.	Round, often rather large and regularly to slightly irregular in outline. Characteristic fairly dense chromatin, nucleolus sometimes present.
Cell shape	With a few processes of variable length.	Cells give off several branches which are extremely thin, often very long and very dense.	Variable, rather large and rounded or, when grouped, somewhat compressed. Sometimes with a few large, rather blunt processes.
Mitochondria	Rarely present: if so, very small with few irregularly arranged cristae.	Few, medium size.	Very few, small with irregularly arranged cristae.
Golgi complexes	Sometimes present, formed of a stack of parallel flattened sacks with associated small vesicles and larger vacuoles.	None found.	Present, positioned close to nucleus, formed from a few stacks of parallel sacks and large number of small vesicles.
Ribosomes	Often present, free or membrane-bound in granular endoplasmic reticulum.	Free and membrane bound, closely packed.	Very many, closely packed on membrane or free, arranged into small groups.
Cytoplasmic inclusions	Numerous, very variable in both number and type. No pigment granules.	Granules of various types, include pigment granules.	Numerous, various sizes and types, some pigment granules.

hoplonemerteans have mucus-secreting submuscular glands embedded in the general body parenchyma. These are most commonly confined to the ventral side, opening to the surface directly via individual ducts. It is presumed that these glands are homologous with the ventral mucous glands found in turbellarian flatworms.

Proboscis apparatus

The proboscis and its associated structures form a characteristic system of the nemertean body. Three distinct components may be recognised, these being the rhynchodaeum, the rhynchocoel or proboscis sheath, and the proboscis itself (Wijnhoff, 1914; Coe, 1943) (Fig. 6).

The proboscis pore opens immediately into the rhynchodaeum, a rather tubular chamber that in many pelagic hoplonemerteans is very short. A rhynchodaeum is missing from the atypical bdellonemerteans. In most of the hoplonemerteans the gut opens ventrally into the posterior end of the rhynchodaeum, so that the anterior pore functions both as a proboscis aperture and a mouth. In the polystyliferous forms the mouth and proboscis openings are either separate or become united and open into a short atrium (Coe, 1943). The rhynchodaeum is a somewhat cone-shaped chamber, widening posteriorly, which extends no further backward than the front of the cerebral ganglia, from which point the proboscis itself starts. The chamber is lined with an epithelium composed of columnar to cuboidal cells, of similar appearance to the epidermis except that cilia may be absent. The occurrence of gland cells in this region depends upon the species; Coe (1905a) reports that in certain *Carinella* (= *Tubulanus*) the anterior rhynchodaeum is provided with abundant compound glands, similar to those of the epidermis, and Gontcharoff and Bierne (1962) demonstrated in *Lineus ruber* the presence of RNA-rich basophilic cells at the base of the rhynchodaeum and in the angle of insertion of the proboscis. Rhynchodaeal musculature, if present, is usually weak and predominantly composed of circular fibres. In a few species of *Prostoma* longitudinal fibres are particularly evident (Gibson and Young, 1971). There is usually a sphincter of circular fibres marking the junction of the proboscis with the rhynchodaeum, this often forming part of the precerebral transverse septum previously mentioned. A similar sphincter is repeated at the oesophageal junction in those forms in which a separate mouth is missing.

The proboscis is a long muscular tubular structure, formed by an invagination of the anterior end of the body. Its muscle layers are therefore homologous with those of the body wall. It is anteriorly

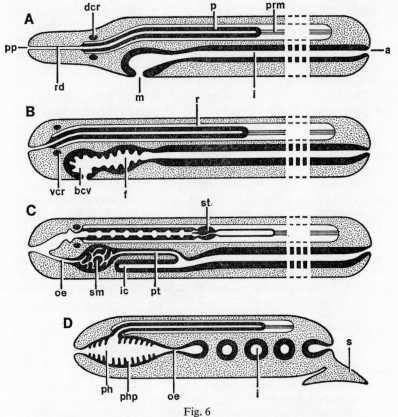

Fig. 6

Schematic vertical longitudinal sections to show the relationships of the gut and proboscis apparatus within the different orders. A, Palaeonemertea (Cephalothricidae); B, Heteronemertea; C, Hoplonemertea (Monostylifera); D, Bdellonemertea. a, anus; bcv, buccal cavity; dcr, dorsal cerebral commissure; f, foregut; i, intestine; ic, intestinal caecum; m, mouth; oe, oesophagus; p, proboscis; ph, pharynx; php, pharyngeal papillae; pp, proboscis pore; prm, proboscis retractor muscle; pt, pyloric tube; r, rhynchocoel; rd, rhynchodaeum; s, sucker; sm, stomach; st, stylet apparatus; vcr, ventral cerebral commissure. (A–C redrawn from Jennings and Gibson, 1969; D redrawn from Gibson and Jennings, 1969)

attached to the tissues of the head in the vicinity of the brain, from where it extends posteriorly in an enclosed cavity, the rhynchocoel. At its posterior extremity it is blind-ending. In length it varies from very short (*Carcinonemertes*, *Gononemertes*) up to two or three times the body length, lying more or less coiled within the rhynchocoel (Bergendal, 1900; Coe, 1902a, 1943; Kirsteuer, 1966).

The two distinct types of proboscis found in the phylum are differentiated as the unarmed form, characteristic of palaeo- and heteronemerteans, and the armed form, representative of the hoplonemerteans. In bdellonemerteans the proboscis, although unarmed, is morphologically similar to that of hoplonemerteans and is generally regarded as being derived from this type.

The unarmed proboscis is essentially a simple tube gradually narrowing to its blind posterior end. In the Palaeonemertea the wall of the proboscis is composed of several recognisable layers (Figs. 7A,C). These are an outer[1] epithelial layer of columnar cells lining the central lumen, and overlying a zone of delicate connective tissue in which the nerve plexus, consisting of two proboscidial nerves, is embedded. The nerves may also run within the epithelium in some species. Beneath the connective tissue are two strong muscle layers

---

Fig. 7

Proboscis structure. **A**, Transverse section of proboscis of *Tubulanus rhabdotus*; **B**, Transverse section of proboscis of *Lineus nigricans*; **C**, Schematic longitudinal section through portion of proboscis of *Cephalothrix bioculata*; **D**, Proboscis barb of *Cephalothrix bioculata*; **E**, Proboscis barb of *Cephalothrix linearis*; **F**, Schematic longitudinal section through part of anterior proboscis of *Paranemertes peregrina*; **G**, Transverse section of anterior proboscis of *Itanemertes nonatoi*; **H**, Transverse section of anterior proboscis of *Pelagonemertes joubini*. **agc**, acidophilic gland cell; **bcl**, basophilic columnar cell; **grc**, granular cell; **icm**, inner circular muscle layer; **ilm**, inner longitudinal muscle layer; **ln**, lateral nerve cord; **lnp**, lateral nerve plexus; **mc**, muscle cross; **ocm**, outer circular muscle layer; **olm**, outer longitudinal muscle layer; **pe**, proboscis epithelium; **pen**, proboscis endothelium; **pn**, proboscis nerve; **ppn**, primary proboscis nerve; **rnp**, radial nerve plexus; **spn**, secondary proboscis nerve; **vc**, vacuolate cell. (A after Corrêa, 1954; B after Corrêa, 1956; C–E redrawn from Jennings and Gibson, 1969; F redrawn from Gibson, 1970; G after Corrêa, 1957; H after Coe, 1926)

---

[1] The relationships of the various proboscis layers in all four orders are reversed when the organ is everted. The terms 'outer' and 'inner' have both been applied to each of the two boundary zones at one time or another; the terms as employed in this text follow the convention used by Coe (1943) so that the epithelial layer that will form the external zone in the everted position is defined as 'outer'. When the proboscis is retracted, therefore, the outer epithelium comes to lie on the inner side of the organ and borders the proboscis lumen.

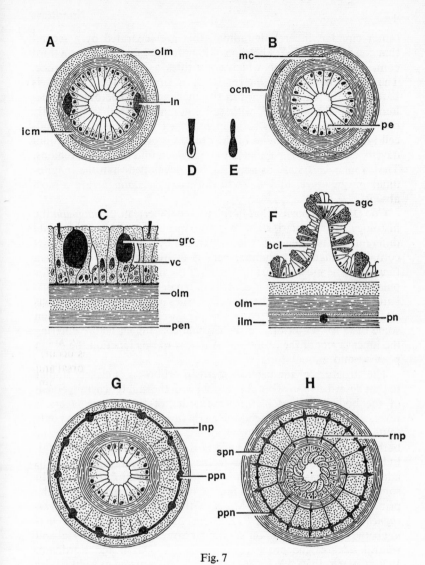

Fig. 7

(outer circular, inner longitudinal) that are separated by a second thin connective region. The zone of longitudinal fibres adjoins a third connective tissue region which constitutes a basement membrane. The organ is bounded internally by a thin endothelium that is bathed by the fluid contained in the rhynchocoel. The outer epithelium is mostly unciliated, with tall columnar cells that frequently contain pigment granules being interspersed with a variety of gland-cell types. The anterior end of the proboscis often has its epithelium developed into raised papillae, or else divided into rhomboid regions. The circular layers are usually less well developed than the longitudinal musculature, and in some forms (*Cephalothrix*) are almost absent.

Coe (1905a) reported that in the palaeonemertean *Carinomella* the histology of the anterior and posterior proboscis chambers is quite different, the two regions being separated by a middle bulbous part containing flagellate columnar cells that are apparently sensory in function. The muscular arrangement of the organ is also unusual for palaeonemerteans in that it consists of a pair of strong lateral muscle bands, originating from the cephalic tissues near the site of proboscis attachment, that do not enter the proboscis immediately but run some distance posteriorly in the rhynchocoel lumen before entering the inner layers of the proboscis. Anterior to this junction the organ possesses no musculature.

The structure of the heteronemertean proboscis is closely similar to that described above but, in keeping with the different arrangement of the body-wall musculature, the order of the muscle layers is reversed. In the family Baseodiscidae there are two muscle layers only (outer longitudinal, inner circular), but in most of the Lineidae the body-wall arrangement is reflected in the addition of an extra longitudinal layer lying just external to the circular layers (Fig. 7B). This additional layer is absent from *Lineus flavescens* and *L. rubescens*. In some forms an extremely thin inner circular layer appears inside the principal longitudinal zone. Many of the lineids (*Dushia, Gorgonorhynchus, Hinumanemertes, Pussylineus*) have muscle crosses occurring between the inner and outer circular layers on both the dorsal and ventral sides of the proboscis (Coe, 1905a; Dakin and Fordham, 1936; Corrêa, 1956, 1963, 1964; Iwata, 1970), situated at right angles to the paired nerves (Fig. 7B). In the lineid genus *Flaminga* muscle crosses occur only on the dorsal side (Corrêa, 1958). The pair of main nerves and any additional nerve fibres lie between the outer longitudinal and outer circular musculature. Amongst the Lineidae the two principal nerves are nearly always only recognisable anteriorly, since they branch profusely to form a complex plexus. The remaining proboscis layers present are as described for palaeo-nemerteans.

Many of the epithelial cells contain rod-shaped secretions, similar in appearance to the epidermal rhabdites of both nemerteans and turbellarians. In some species they possess characteristic shapes and seem to be contained within a basal cup (Figs. 7D,E). They may be solitary or arranged together in small numbers, as in *Cephalothrix* species (Jennings and Gibson, 1969) (Fig. 7C), or be aggregated into such large numbers that they produce a 'pincushion' effect, found in the heteronemertean *Lineus ruber* (Gontcharoff, 1957; Bierne, 1962a; Ling, 1971). Similar structures have been recorded from *Carinomella lactea* (Coe, 1905a), *Uchidana parasita* (Iwata, 1967) and *Hinuma-nemertes kikuchii* (Iwata, 1970). Iwata rather confusingly calls them stylets.

The possible role of these structures, which can apparently be projected from the proboscis surface when the organ is everted, is interpreted as being analogous to that of the hoplonemertean central stylets in that they could cause a wound in the body of a potential prey, through which toxic or other secretions could be poured. Whether the structures themselves contain soluble toxins, or whether these secretions originate from other epithelial regions, is not yet known. The structures are lost after use and can be replaced by new ones synthesised proximally by their containing cells. Whatever their other functions, they clearly assist the proboscis in its mechanical hold on the prey.

In some species of *Cerebratulus*, *Lineus* and *Micrura* the anterior proboscis epithelium bears cells containing urticating filaments (Coe, 1905a). These have a similar appearance to the nematocysts of coelenterates; Coe states that the secretions of the cells are extremely viscid and are presumed to assist the proboscis in its grip. The cells are usually confined to two longitudinal welts, and their origin is not known.

Considerable interspecific variation occurs with respect to the precise arrangement of the proboscis layers, and differences can also be found between different parts of the same organ. The most common modification is a reduction of the anterior musculature.

In the hoplonemerteans the proboscis musculature comprises inner and outer circular layers enclosing a very much thicker longitudinal zone. The outer circular region is often missing posteriorly.

Each species possesses a number of longitudinal proboscidial nerves, situated near the periphery of the longitudinal muscle zone and connected laterally by a nerve plexus that can be so well developed as to effectively divide the longitudinal musculature into inner

44   Nemerteans

and outer portions (Figs. 7G,H). The precise number of nerves is not
always constant for a given species, and may be quite variable. In
the terrestrial hoplonemertean *Geonemertes australiensis*, fourteen is
the commonest number, but from eleven to twenty-one have been
recorded (Hickman, 1963). Conversely, in other forms the number
is so inflexible as to be of diagnostic value.

Amongst the polystyliferous hoplonemerteans, some species of the
reptantic genus *Drepanophorus* may have more than thirty nerves. In
several bathypelagic forms larger (primary) and smaller (secondary)
proboscis nerves are present, and alternate regularly, the secondaries
being placed midway between the primaries in the lateral nerve
plexus (Fig. 7H). *Planktonemertes agassizii* has twenty-five nerves of
each type, *Pelagonemertes joubini* fifteen. The number of secondaries
is sometimes quite variable, *Plionemertes plana*, for example, having
twenty-four primary nerves but only nineteen to twenty-three secondaries in some parts of the proboscis, but a full twenty-four in others
(Coe, 1926).

The hoplonemertean proboscis is divisible into three distinct
regions, an anterior thick-walled tube, a short bulbous muscular
middle region (the stylet bulb), and a posterior blind-ending tube
that is often shorter and much more slender than the anterior portion. Anterior and posterior chambers are joined by a narrow canal
that penetrates the muscular septum or diaphragm of the stylet bulb
(Figs. 8A,B). The stylet-bulb region is better differentiated in the
monostyliferous than the pelagic polystyliferous forms (Coe, 1926).

The anterior epithelium is arranged into flattened, conical, scale-
like or filamentous papillae composed of tall gland cells and elongate
columnar cells (Fig. 7F). Glands are filled with an acidophilic pro-

Fig. 8

Proboscis structure. **A**, Schematic representation of a hoplonemertean
proboscis in retracted position; **B**, as A, but protruded; **C**, Stylet
apparatus of *Amphiporus californicus*; **D**, Stylet apparatus of *Amphiporus formidabilis*; **E**, Central proboscis bulb of *Prostoma graecense*;
F, Proboscis of *Carcinonemertes carcinophila*; **G–J**, Central stylet and
basis of *Emplectonema gracile* (**G**), *Amphiporus lactifloreus* (**H**), *Prostomatella enteroplecta* (**I**) and *Paranemertes peregrina* (**J**). **ape**, anterior
proboscis epithelium; **asp**, accessory stylet pouch; **ast**, accessory stylet;
cst, central stylet; **gc**, gland cell; **ppe**, posterior proboscis epithelium;
prm, proboscis retractor muscle; **r**, rhynchocoel; **ren**, rhynchocoel
endothelium; **sbc**, stylet bulb chamber; **sbd**, stylet bulb duct; **sbm**, stylet
bulb musculature; **sbn**, stylet bulb nerve; **sbs**, stylet bulb sphincter;
stb, stylet basis. (A, B redrawn from Gibson, 1970; C, D, J redrawn
from Coe, 1905a; E redrawn from Coe, 1943; F redrawn from Coe,
1902a; G, H redrawn from Brunberg, 1964; I redrawn from Corrêa,
1954)

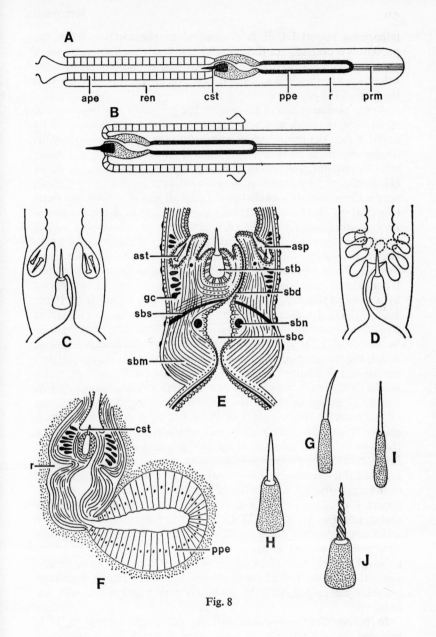

Fig. 8

teinaceous material that is discharged to the exterior when the proboscis is everted, but their precise role is not yet known. Some of the glands contain enzymes (see Chapter 3), but the contents of many others are apparently viscous in nature and are presumed to assist in the grip of the organ.

At the posterior end of this region the proboscis narrows into the stylet bulb, the thick muscular diaphragm of this part housing on its anterior face the central stylet apparatus (Fig. 8E). Monostyliferous hoplonemerteans have a single central or main stylet, a usually smooth straight, sometimes curved (*Emplectonema gracile*), fluted (*Emplectonema purpuratum*) or grooved (*Paranemertes peregrina*), needle-like structure socketed into a conical granular mass, the basis (Figs. 8G–J). There are usually two, sometimes more, lateral or accessory pouches containing two to several reserve or accessory stylets in various stages of development (Figs. 8C–E). There may be up to twelve accessory pouches in *Amphiporus formidabilis*, the most recorded for a monostyliferous hoplonemertean. The middle region is constricted from both the anterior and posterior proboscis portions by a strong sphincter, composed principally of circular fibres (Coe, 1905a, 1943). Its anterior diaphragm, in which the stylet basis is embedded, contains many elongate glands whose secretions seem to be responsible for the synthesis of the basis. Accessory pouches, which may be anterior, lateral or slightly posterior to the main stylet, are formed from a number of large vacuolated cells, each of which secretes a single accessory stylet into its interior. In the Monostylifera this development occurs in the following manner. An invagination of the anterior diaphragm wall gives rise to a pouch into which glandular secretions are poured and accumulate. These give rise to the main basis. An accessory stylet then migrates centrally to become firmly held by the basis socket. The central stylet can thus be readily replaced if it becomes lost or damaged. The number, size and shape of the accessory stylets are often of specific character. In the parasitic genera *Carcinonemertes* has secondarily lost its accessory pouches and stylets (Fig. 8F) (Coe, 1902a,b), whilst in *Gononemertes* the entire stylet apparatus has degenerated (Coe, 1943). The entire proboscis structure is reduced in the genus *Sacconemertella* (Iwata, 1970), although the rhynchocoel extends for the full body length. In this genus the proboscis structure is unusual in that the muscle layers are irregularly arranged, and the proboscis nerves and stylet basis are indefinite.

In polystyliferous hoplonemerteans the single central stylet is replaced by a sickle-shaped basis or pad which bears numerous minute top-shaped central stylets (Figs. 9A–D). These, like the stylets

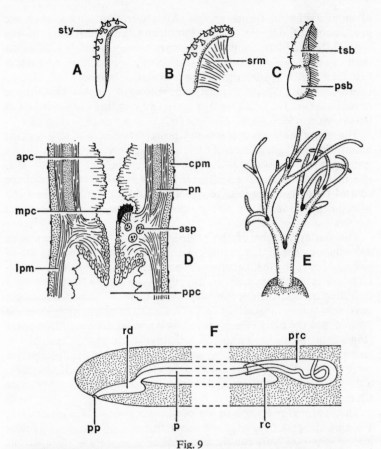

Fig. 9

Proboscis structure. **A–C**, Central stylets and basis of *Plionemertes plana* (**A**), *Planktonemertes agassizii* (**B**) and *Pelagonemertes brinkmanni* (**C**); **D**, Longitudinal section through stylet region of *Nectonemertes pelagica*; **E**, Branched proboscis of *Gorgonorhynchus repens* in everted position (black regions represent positions of valves); **F**, The arrangement of the rhynchocoel and proboscis of *Tubulanus borealis*. **apc**, anterior proboscis chamber; **asp**, accessory stylet pouch; **cpm**, circular proboscis musculature; **lpm**, longitudinal proboscis musculature; **mpc**, middle proboscis chamber; **p**, proboscis; **pn**, proboscis nerve; **pp**, proboscis pore; **ppc**, posterior proboscis chamber; **prc**, posterior rhynchocoel chamber; **psb**, proximal stylet basis region; **rc**, rhynchocoel caecum; **rd**, rhynchodaeum; **srm**, stylet radial musculature; **sty**, stylet; **tsb**, terminal stylet basis region. (A–D redrawn from Coe, 1926; E after Dakin and Fordham, 1931; F redrawn from Friedrich, 1936)

of monostyliferous forms, are glandular secretions and as such are presumably of organic nature. Paralleling the multiplicity of the central stylets, there are usually more accessory pouches and stylets than in other groups, although some pelagic species have apparently lost their stylet apparatus. *Pelagonemertes brinkmanni* has nine to twelve accessory pouches, whilst some species of *Drepanophorus* have more than twenty, with a total of up to two hundred accessory stylets (Coe, 1905a, 1926).

The stylet bulb has a thick and powerful muscular wall formed mainly from longitudinal and oblique fibres, developed into a strong sphincter at each end and lined by a tall epithelium (Fig. 8E). It is this musculature that is responsible for the penetrative power of the central stylet as an aggressive structure, and for pumping the toxic secretions, produced by the posterior proboscis, into the stylet wound.

The posterior tube of the armed proboscis has a more simple columnar epithelium, normally without papillae, numerous gland cells not restricted to an acidophilic nature, and a diminishing musculature. Secretions produced by this region are clearly the toxins used in killing or inactivating live prey, since from the functional aspect only substances originating from this part of the organ could be poured into the body of a potential food animal via the stylet wound (Figs. 8A,B) (Gibson, 1970). This theoretical interpretation of the secretions' function is to some extent supported by the distribution recorded by Kem (1971) for the Pacific *Paranemertes peregrina*, although larger quantities of the toxin do occur in other tissues (see Chapter 3).

The unarmed proboscis of the bdellonemerteans (*Malacobdella*) is a simple elongate tube as in the other unarmed groups, but its histological structure, particularly with respect to its muscle arrangement and papilla formation, is as described for hoplonemerteans (Wijnhoff, 1914; Riepen, 1933).

A most unusual type of proboscis, characteristic of the genus *Gorgonorhynchus*, is in its basic anatomy comparable with the typical heteronemertean form but differs in that it branches dichotomously into sixteen or thirty-two proboscides, and has a development of luminar valves associated with the extrusion of a complex organ (Dakin and Fordham, 1931, 1936) (Fig. 9E).

In hoplonemerteans the stylet armature when first formed in juvenile worms is of smaller size than found in adults, and more or less commensurate with body size. As the nemerteans grow and increase their bulk, so the occasional replacement of both stylet and basis keeps pace and eventually gives rise to the normal adult form.

The mechanisms involved are essentially the same as previously described. If, however, the species concerned is one in which the adults are small (*Prostoma*, *Tetrastemma*), the initial stylet apparatus may be of an appropriate size and the only replacement that takes place is that normally involved following stylet loss.

The premature liberation of an accessory stylet has been reported for *Amphiporus pulcher* (Coe, 1943). When this happens the stylet becomes lodged in the basis pouch before this is fully formed, with the result that a double styletted basis is produced when the later, normal arming takes place. The premature stylet is often lodged in the posterior part of the basis where it presumably will either remain or degenerate. A similar, but more complex, anomaly was found in a specimen of *Tetrastemma candidum* by Caullery (1908), which had a double stylet bulb region, the second one being positioned at the rear end of the proboscis and orientated posteriorly. Both bulb areas were fully armed.

The proboscis lies free within the rhynchocoel, attached by its posterior end via a retractor muscle (Figs. 6, 8A,B). Such a muscle is absent from some nemerteans (*Carcinonemertes*, *Cerebratulus lacteus*, *C. melanops*, *Gorgonorhynchus*, *Zygeupolia*, and certain pelagic Polystylifera), the proboscis then being posteriorly unattached (Thompson, 1901; Coe and Kunkel, 1903; Coe, 1895, 1926). The fibres which make up the retractor muscle are derived from the longitudinal proboscis musculature, and may join the rhynchocoel epithelium either at its posterior extremity or some way anterior to this. In the genus *Carcinonemertes* the proboscis sheath is reduced to the merest rudiments, and the posterior proboscis chamber is embedded in the general parenchyma (Coe, 1902a,b; Kirsteuer, 1966).

The rhynchocoel is closed at both ends, anteriorly by the rear face of the rhynchodaeum to which the front of the proboscis is attached, posteriorly by a simple sealing of the tubular structure. It is lined by a flattened endothelium similar to that covering the inner surface of the proboscis, resting on a thick or thin basement layer of homogeneous connective tissue. In several species of *Cerebratulus* and other forms the endothelium contains considerable numbers of gland cells. These are commonly arranged in irregular rows positioned close to the rhynchocoel vessels, and discharge their contents into the rhynchocoel lumen. This, in all species, is filled with a fluid in which float various granular amoeboid discs, globules and spindle-shaped cells, presumably originating from the rhynchocoel endothelium. The function of these inclusions is not known but they may well be involved in some way with the nourishment of the proboscis apparatus, particularly since this structure lacks any development of blood vessels. Most of the inclusions possess some power of amoeboid locomotion, and several can be found forming slender pseudopodial processes. Often several discs become aggregated into larger balls. The fluid of the rhynchocoel is usually colourless, but is pale red in *Paranemertes*. In *Amphiporus flavescens* the amoeboid discs are red or yellow (Coe, 1905a).

The wall of the proboscis sheath consists of an outer endothelium formed from connective tissue and muscle fibres. There are usually three muscle layers, arranged as outer longitudinal, middle circular and inner longitudinal zones, but in several species the layers are far from clear. In *Drepanophorus* and some *Amphiporus* species distinct muscle layers are absent from the middle and posterior portions, the regions consisting instead of a network of interwoven longitudinal and circular fibres. The rhynchocoel wall of bathypelagic species is massive compared with the size of the fragile epidermis and reduced body-wall musculature (Coe, 1926).

The rhynchocoel is in most nemerteans shorter than the body, extending half to three-quarters of the length, but in others may be as long as the body (*Geonemertes, Potamonemertes*) or not extend beyond the anterior third.

The reptantic Polystylifera (*Drepanophorus, Siboganemertes, Uniporus*) are the only group with paired rhynchocoel diverticula, serially repeated lateral pouches, which may be branched, replicating the diverticula of the intestine. They may extend ventrally below the intestine well towards the mid-line. They possess special muscles associated with them (Coe, 1905a; Stiasny-Wijnhoff, 1923, 1926). Other species have one or more middorsal or midventral pouches; the rhynchocoel of *Uchidana parasita*, for example, protrudes ventrally in the anterior cerebral region to form a short compressed diverticulum lined with a typical endothelial membrane (Iwata, 1967).

In the palaeonemertean *Tubulanus* (*Carinella*) *frenata* and other related species the rhynchocoel is enlarged in the nephridial region to form a very large chamber in which the greater portion of the proboscis lies coiled. Behind this level the rhynchocoel is contracted into a narrow tube. The large anterior chamber is continued posteriorly and ventrally as a short, blind-ending caecum, lying below the main chamber which houses the proboscis (Coe, 1905a). A most unusual rhynchocoel is found in *Tubulanus holorhynchocoelomicus* (Friedrich, 1958) and *T. borealis* (Friedrich, 1936), where it consists of two distinct chambers joined by a narrow canal arising from the dorsal wall of the anterior chamber (Fig. 9F). The proboscis in these forms is contained entirely within the anterior region.

The musculature of the proboscis apparatus may be considerably modified from that of the body wall, although essentially replicating it. The rhynchocoel is fundamentally a simple split within the musculature that divides into two major portions. One portion forms the proboscis muscles, the other corresponds to the proboscis sheath. For this reason the rhynchocoel cavity is strictly schizocoelic, but in being involved only in such a specialised and peculiar body arrangement does not destroy the essentially acoelomate nature of nemerteans as a group.

The proboscis is used principally in the capture of live prey, but may also be used as a defensive weapon or, as in some terrestrial and freshwater species, as a locomotory organ. It is everted rapidly and explosively by hydrostatic pressure created by muscular contractions squeezing the rhynchocoel fluid. Since the anterior end of the proboscis is attached to the inner margins of the rhynchodaeum, it necessarily turns inside out in a way that can be easily demonstrated by a glove finger. This action brings the glandular and often papillar lining to the exterior, the secretions being employed as discussed earlier. In armed forms eversion occurs only as far as is necessary to bring the central stylet to a terminal position, where the armature can then be used (Fig. 8B), although there is no reason to suppose that even in these nemerteans further eversion cannot take place. Proboscis withdrawal is effected by the retractor muscles or, in species which do not possess one, by other musculature of the proboscis apparatus, concurrent with a release of hydrostatic pressure.

Proboscis ejection can be accurately controlled by many nemerteans. In *Paranemertes peregrina* it can be slowly and only partially everted or retracted whilst apparently being used as an exploratory organ (Roe, 1970). In other species it is withdrawn completely after use and is either not used again until a later time, or is fully everted immediately and as rapidly as before. Repeated proboscis eversion is frequently seen in instances in which the potential prey is of a comparatively large size.

Nervous system

The nervous system of nemerteans consists of a pair of cerebral ganglia, a pair of large ganglionated nerve cords, a main nerve plexus of which the lateral nerves comprise the principal components, and often several minor or subsidiary plexi. The position of the central nervous system relative to the body-wall layers is of major taxonomic value within the phylum (Fig. 5). The greatest variation is found amongst the palaeonemerteans; in some genera (*Carinina*) the cerebral ganglia and nerve cords are situated within the epidermis, in most they are found within the dermis (*Tubulanus*), and in a few they occur at least in part (posterior to the oesophagus) in the body-wall musculature (*Carinoma, Carinomella*). In *Cephalothrix* the lateral nerves run the full body length in the longitudinal muscles (Hubrecht, 1880a; Hylbom, 1957). In the heteronemerteans the main nerve cords run in the circular muscle layers, the cerebral ganglia being closely surrounded by the cephalic muscles, whereas in hoplo- and bdellonemerteans the nervous system is placed internal to the body wall and lies within the parenchyma. This inward migration of the system approximately parallels the increase in body complexity found within the phylum.

In many species the nerves in living specimens are tinged red or yellowish, this colouration being particularly evident in many species of *Amphiporus* and *Prostoma*.

The cerebral ganglia are usually four-lobed, except in bdellonemerteans where only a single ganglion occurs on either side of the body (Riepen, 1933), arranged as dorsal and ventral lobes on either side of the mid-line (Figs. 10A–D). The dorsal lobe is larger than the ventral in palaeonemerteans, the converse is true in hetero- and hoplonemerteans (Coe, 1927b). In the pelagic groups the two lobes on either side are very closely fused together (Coe, 1926). The ganglia are usually ovoid in shape, but some variation is found in certain forms. In *Uchidana* the dorsal lobes are posteriorly bifurcated (Iwata, 1967), and in *Gorgonorhynchus* the entire brain arrangement is somewhat unusual (Dakin and Fordham, 1936) (Fig. 10E).

Lateral junction between ganglia is achieved by dorsal and ventral cerebral commissures situated above and below the rhynchocoel respectively, with the ventral connective frequently being the larger. The cerebral ganglia complex thus encircles the anterior portion of

---

Fig. 10

Central nervous sytem. A–E, Cerebral ganglia and principal nerve roots of *Tubulanus annulatus* (A), *Cerebratulus marginatus* (B), *Prostoma jenningsi* (C), *Malacobdella grossa* (D) and *Gorgonorhynchus repens* (E); F, Nervous system of *Neuronemertes aurantiaca*. cdn, caudal nerves; chn, cephalic nerves; co, cerebral organ; dcl, dorsal cerebral lobe; dcr, dorsal cerebral commissure; dln, dorsolateral nerve; dn, dorsal nerve; dpn, dorsal peripheral nerve; gn, gastric nerve; ln, lateral nerve cord; lpn, lateral peripheral nerve; mg, metameric ganglia; oen, oesophageal nerve; pc, posterior (anal) commissure; vcl, ventral cerebral lobe; vcr, ventral cerebral commissure; vpn, ventral peripheral nerve. (A, B redrawn from Bürger, 1897–1907; D redrawn from Riepen, 1933; E redrawn from Dakin and Fordham, 1936; F redrawn from Coe, 1927b)

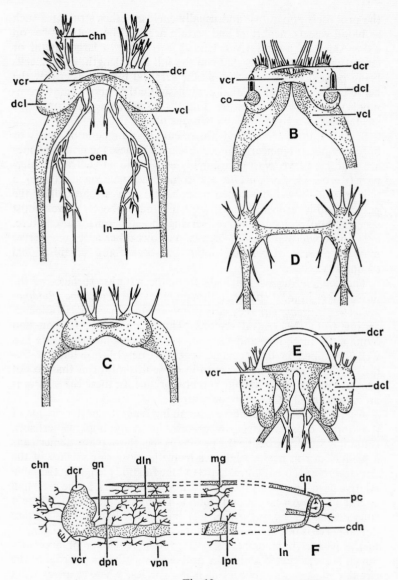

Fig. 10

the proboscis apparatus, and usually encloses other structures such as blood vessels, nephridia and certain nerves. The ventral lobes on either side are posteriorly extended into the two large lateral or lateroventral nerve cords that run the full body length, ganglia cells being present in these as well as the cerebral lobes. Posteriorly the nerve cords unite via an anal commissure either above (most commonly) or below the intestine (Fig. 10F). The nerves are connected at intervals along the body by nervous plexi.

Several minor nerves that are essentially ganglionated cords or fibres are present in nearly all nemerteans. Of these the principal ones are a dorsal nerve, with occasionally an additional ventral counterpart (*Carinoma, Carinomella*, and some other palaeonemerteans), a pair of usually well-developed oesophageal nerves (particularly the palaeo- and heteronemerteans), cephalic and proboscis nerves (most species), and peripheral nerves serving the epidermis, musculature, eyes and epidermal sensory organs. Various other accessory nerves may be present, depending upon the species and organisational complexity.

The cephalic nerves originate from the anterior surfaces of the dorsal lobes in most species, although in several (*Prostoma, Potamonemertes, Zygeupolia*) they are supplemented by similar branches arising from the ventral regions (Thompson, 1901; Gibson and Young, 1971; Moore and Gibson, 1973). Cephalic nerves are less well developed in bathypelagic species (Coe, 1927a), but are frequently large and numerous even in those littoral forms that do not possess either eyes or definite cerebral organs. In these cases there is an indication of other cephalic senses.

A pair of dorsolateral nerves, stemming from the posterior side of the dorsal cerebral lobes, is present in many hoplonemerteans, running posteriorly beside the proboscis sheath. In many nemerteans a median dorsal nerve originates from the posterior surface of the dorsal cerebral commissure, to extend the full body length in the same relative position, with respect to the body-wall layers, as the principal lateral nerves. An exception is found in the Hoplonemertea, where it runs nearer the body surface, just inside the dermis. The median dorsal nerve is not in direct communication with the cerebral ganglia in the pelagic hoplonemerteans (Fig. 10F). An inner or accessory dorsal nerve, situated above the proboscis sheath and internal to the circular musculature, occurs in many species, where it arises from branches of the outer median dorsal nerve and not from the brain directly. This nerve serves the rhynchocoel.

A pair of oesophageal or foregut nerves (the gastric nerves in pelagic species) arises either from the two ventral cerebral lobes or from the ventral commissure. These extend to the foregut, and then run posteriorly within its walls. Transverse connections join the two nerves at various places.

A median ventral nerve, situated in the body-wall musculature, occurs in some palaeonemerteans.

The proboscis nervous supply is simplest in the anoplan groups. In these forms a single pair of lateral proboscis nerves (Fig. 7A) usually arise from the ventral cerebral commissure and extend posteriorly in the organ either immediately beneath the epithelium or between the muscle layers. The two nerves branch to form a cylindrical plexus within the proboscis wall, less profusely in the palaeo- than in the heteronemerteans, where the plexus formation may be so complex as to obscure the basic bilateral nature of the system.

Conversely, in the armed enoplans, the number of proboscis nerves ranges from seven to fifty, these arising either from the anterior face of the brain, where the ventral commissure joins the ventral lobes, or by the dichotomous branching of a single pair of nerves originating from the same point. The nerves are more or less regularly disposed within the proboscis wall, running anteriorly in the longitudinal muscle layer (Figs. 7G,H), posteriorly below the lining epithelium. In the stylet bulb region the nerves are fused into an irregular plexus that is sometimes double. The nerves of the anterior proboscis are laterally joined by a fibre network that forms the proboscis nerve plexus, effectively dividing the longitudinal musculature into two layers. Radial fibres may extend from the plexus to form additional plexi of which one is positioned beneath the epithelium, the other inside the outer band of circular muscle fibres (Fig. 7H).

In anoplan nemerteans the main lateral nerves are connected with each other and with the median dorsal nerve by the body-wall nerve layers, formed by anastomosing fibres arranged more or less irregularly into a peripheral network. In heteronemerteans this lies between the circular and outer longitudinal muscle layers, consisting of a cylindrical nervous plexus extending most of the body length. A secondary plexus may lie internal to the circular musculature, formed from branches of the main plexus passing through the circular fibres. In *Tubulanus* and many other palaeonemerteans a delicate nerve plexus lies between the epidermis basement lining and the outer circular muscle layer. The plexus of *Hubrechtia* is massive. In all cases the nervous layers consist of very delicate nerve fibrils and a few small nerve cells supported by a framework of connective tissue.

The hoplonemertean nervous system is best known for its development in the bathypelagic forms (Coe, 1926, 1927b), where additional nerves are of quite common occurrence (Fig. 10F). These include a pair of small gastric nerves, serving the stomach in those species with a shortened gut, that are often supplemented by extra nervous fibrils originating from the ventral cerebral commissure. In some forms a pair of small nerves leaves the lateral nerve cords and enters the anterior region of the pyloric tube. Many species have a pair of nerves in the anterior region, originating from the dorsal cerebral lobes, passing posteriorly along the internal border of the body-wall musculature, near the dorsolateral margins of the rhynchocoel. These nerves are joined by delicate fibres with both main dorsal nerves and the dorsal peripheral supply of the lateral nerve cords. Efferent branches serve the body musculature on the dorsal side as well as innervating the rhynchocoel. These nerves are not found in all species, and, if present, are entirely confined to the anterior regions. In other species, and towards the middle of the body in all, they are replaced by fibres originating from the dorsal peripheral nerves.

The peripheral nervous system in pelagic hoplonemerteans consists of a pair of large nerves, one dorsal and one ventral, leaving the lateral nerve cords at each interdiverticular space. From the dorsal branch a third, smaller peripheral nerve – the lateral nerve – branches out. The three nerves between them serve the rhynchocoel, dorsoventral musculature, body wall and intermuscular plexi (Fig. 11A).

In the swimming Polystylifera the caudal nerves serving the posterior end arise in one of two ways. In *Proarmaueria* the posterior end of each lateral nerve divides into two branches, the larger of these joining up with the anal commissure, the smaller innervating the body extremities. However, in *Neuronemertes* and other species the caudal nerves stem directly from the anal commissure and dichotomy of the lateral nerves does not occur. Adult male *Nectonemertes* and both sexes of *Balaenanemertes* have their paired anterior tentacles innervated by development of the lateral peripheral nerves in the appropriate region (Fig. 11B). These so-called large tentacular nerves are directly homologous with the peripheral nerves from other parts of the body.

Many nemerteans have the proboscis sheath supplied with nerves that enter it at the point of attachment of the proboscis, but in others these nerves arise as branches of the dorsal peripheral nerves.

A few nemerteans have an accessory lateral nerve arising on each side of the body from the dorsal cerebral lobes, extending posteriorly, either adjoining or slightly apart from the upper surface of the main lateral cords (Figs. 11C,D). This condition, apparently confined to some of the hoplonemerteans, is found only in most species of *Geonemertes* (Pantin, 1969), the genus *Oerstedia* (Stiasny-Wijnhoff, 1930), the pelagic Polystylifera (Coe, 1926, 1927b) and in some

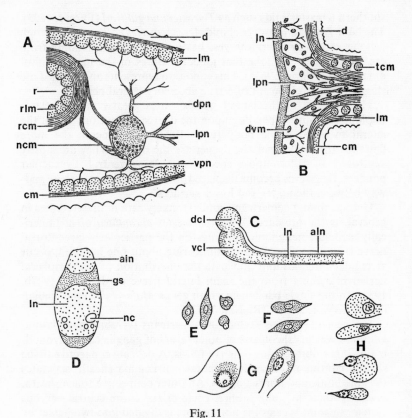

Fig. 11

Central and peripheral nervous systems. A, Portion of the peripheral nervous system of *Pelagonemertes joubini*; B, Horizontal section through base of cephalic tentacle of *Nectonemertes mirabilis*, showing nerves originating from lateral nerve cord; C, Origin and position of accessory lateral nerve cord; D, Transverse section through lateral and accessory lateral nerve cords of *Geonemertes hillii*; E–H, Neurosecretory cells of *Cerebratulus marginatus*, cell types a (E), b (F), c with c_1 on right, c_2 on left (G) and d (H). **aln,** accessory lateral nerve; **cm,** circular musculature; **d,** dermis; **dcl,** dorsal cerebral lobe; **dpn,** dorsal peripheral nerve; **dvm,** dorsoventral muscles; **gs,** ganglionic sheath; **lm,** longitudinal musculature; **ln,** lateral nerve cord; **lpn,** lateral peripheral nerve; **nc,** neurochord cell; **ncm,** nerve cord muscles; **r,** rhynchocoel; **rcm,** rhynchocoel circular musculature; **rlm,** rhynchocoel longitudinal musculature; **tcm,** tentacular circular musculature; **vcl,** ventral cerebral lobe; **vpn,** ventral peripheral nerve. (A, B redrawn from Coe, 1927b; C after Pantin, 1969; D redrawn from Pantin, 1969; E–H redrawn from Bianchi, 1969a)

southern tetrastemmids such as *Tetrastemma gulliveri* (Bürger, 1893). The New Zealand prosorhochmid *Acteonemertes bathamae*, although lacking accessory lateral nerves, has a short nerve extending from each dorsal cerebral lobe that joins the fibres of the main longitudinal cords (Pantin, 1961a). This condition represents an intermediate stage in the development of a true accessory lateral nerve.

The position of the lateral nerves varies greatly in the pelagic hoplonemerteans, depending upon the degree of development of the lateral margins of the body (Coe, 1927b). In general, the more flattened the body form, the deeper are the nerves. They are mostly ventrolateral, but in those species with large ventral diverticular pouches, the nerves become displaced dorsally and lie somewhere midway between the upper and lower surfaces.

Unique among nemerteans, and extremely rare in invertebrates in general, is the presence in *Neuronemertes aurantiaca* of metamerically arranged ganglionic swellings on the particularly large dorsal nerve (Fig. 10F). The ganglia, numbering about one hundred, occur at regular intervals and duplicate the distribution of the peripheral nerves branching from the main lateral nerve cords (Coe, 1927b, 1933). In the single bdellonemertean genus *Malacobdella* the median dorsal nerve is absent (Riepen, 1933).

The central nervous system in nemerteans (cerebral ganglia and lateral nerve cords) consists of outer layers of ganglia cells surrounding an inner fibrous zone of nerve fibrils. A delicate connective tissue sheath, continuous with the cerebral neurilemma sheath, separates the ganglionic and fibrous layers. An outer connective tissue sheath, several μ thick in some species, encloses the entire central nervous system. In some species the fibrous core is divided into two halves by a median layer of ganglion cells (*Proarmaueria pellucida*, *Planonemertes lobata*). Ganglion cells, which are nearly all unipolar, are of several sorts, differing in their shape, position and reaction to histological stains. Bianchi (1969a,b) recognises six nerve-cell types, summarised in Table 3, from the heteronemertean *Cerebratulus* (Figs. 11E–H).

Earlier studies on nerve-cell types were made by Bürger (1895) and Montgomery (1897). These authors, although only recognising four cell categories, closely agree with the differentiation given by Bianchi. Cells equivalent to Bianchi's types *a*, *b* and *c* have been found in *Lineus* by Lechenault (1962, 1963). With the possible exception of the various cephalic nerves, lesser nerves and nervous plexi also contain ganglion cells.

In some hetero- and hoplonemerteans particularly large-sized neurochord cells are also found in the central nervous system. Some

Table 3
Nerve cell types in *Cerebratulus* (from Bianchi, 1969a,b).

Cell type Group	Subgroup	Position	Description
a		Dorsal lobes of brain.	Small ganglia cells, shortened pyriform shape, nuclei spherical, axonal pathways (neurosecretory) run towards neurohemal area, which is in close contact with lateral blood vessel.
	a_1	Mediodorsal, close to cephalic clefts.	Most numerous, in groups of about 30.
	a_2	Throughout dorsal lobes.	Scattered, often in groups of 2 or 3.
b		Ventral lobes of brain and lateral nerve cords.	Differ from type *a* in shape and disposition.
c		Both brain lobes and lateral nerve cords.	Very variable in size.
	c_1		Medium sized, not grouped or rarely so.
	c_2		Very large, never grouped.
d		Dorsal brain lobes.	Differ from *c* types in staining affinities, very variable in size, sometimes grouped.

genera have only a single pair present in the cerebral ganglia, whilst others have several pairs. The single large neurochords (nerve processes) extend from these cells into the lateral nerve cords which may contain additional, often numerous, cells of the same type. When present in the lateral nerve cords they are very obvious on account of their enormous size (Fig. 11D). If a number are present together, they become aggregated into distinctive tracts or bundles. Neurochord cells have not been found in the pelagic hoplonemerteans (Coe, 1927b).

The presence of such features as the peripheral ganglionated plexus and other longitudinal cords is strong evidence that the nervous system is derived from a radial arrangement, according to Hyman (1951).

Sense organs

All nemerteans possess certain highly specialised sense organs, including eyes, oblique and horizontal cephalic grooves or furrows, cerebral, lateral and frontal sense organs, and sensory epithelial cells.

The sensory epithelial cells, rather slender structures that terminate distally in a single projecting cilium or seta (Fig. 4A) are known from particular body regions in some species (*Ototyphlonemertes, Prostoma*), where they are especially prevalent in the extreme anterior and posterior extremities (Böhmig, 1898; Corrêa, 1951). These cells, which presumably have a tactile function, are not easy to see even in living specimens, and may well be of much more general occurrence within the phylum.

In the epidermis of some pelagic hoplonemerteans (*Cuneonemertes*) there occur sensory pits or depressions, the so-called integumentary sense organs (Coe, 1926, 1927a). These are composed of clusters of sensory nerve cells interspersed with supporting cells, arranged into a rather flask-shaped organ (Fig. 12A). They are especially common on the head, occurring less frequently over other regions of the body. Similar structures are also found in several palaeonemertean genera (*Carinoma, Carinomella, Tubulanus*) where they appear as a number (usually six to twelve) of sensory pits arranged in a row on the dorsal side of the head (Coe, 1905a). Slightly different sensory organs, although possibly homologous, are found in other palaeonemerteans. In these a single pit or lateral sense organ occurs on either side of the body near the excretory pore. The organ is composed of attenuated epithelial cells that are underlain by nervous material, the whole structure being protrusible by means of a specifically associated mus-

---

Fig. 12

Sensory organs. **A**, Integumentary sense organ and **B**, subepidermal sense organ of *Cuneonemertes gracilis*; **C**, Frontal organ and cephalic glands of geonemerteans; **D**, Section through eye of *Drepanophorus spectabilis*; **E, F**, Otoliths and statocysts of *Ototyphlonemertes lactea*; **G**, Transverse section through lateral sense organ of *Carinella* (= *Tubulanus*) *frenata*; **H**, Position of cerebral organs relative to cerebral ganglia in *Lineus ruber*; **I**, Schematic drawing of cerebral organ in horizontal section of *Lineus ruber*. **bnc**, bipolar nerve cells; **c**, cerebral organ canal; **cg**, cerebral ganglia; **cgl**, cephalic gland; **cm**, circular musculature; **co**, cerebral organ; **coo**, cerebral organ opening; **cs**, cephalic slit; **d**, dermis; **ec**, eye capsule; **egc**, 'extra-ganglionic canal'; **en**, nerve of eye; **ep**, epidermis; **fo**, frontal organ; **fvc**, fibrous visual cell; **gla**, gland cell type *a*; **glb**, gland cell type *b*; **lm**, longitudinal musculature; **ln**, lateral nerve cord; **lso**, lateral sense organ; **ndc**, nodular cell; **nv**, nerve; **pgc**, pigmented visual cell; **sc**, sensory cilia; **spb**, secretory pool of basal cells; **stt**, statocyst; **vsc**, vesicular cell. (A, B redrawn from Coe, 1926; C after Pantin, 1969; D redrawn from Bürger, 1897–1907; E, F redrawn from Corrêa, 1954; G after Coe, 1905a; H, I redrawn from Ling, 1969b)

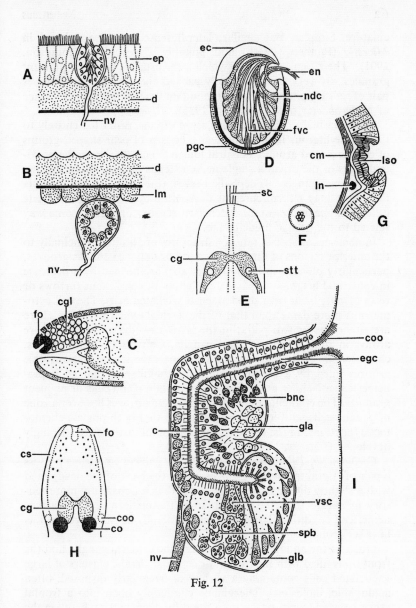

Fig. 12

culature. Simpler, but similar, lateral sense organs are found in *Micrella* (Punnett, 1901b) and possibly *Zygeupolia* (Thompson, 1901). Their sensory cells, which are usually without pigment granules, are very slender, closely packed and possess long cilia. A pair of sensory pits are similarly found on the anterior dorsal surface of *Malacobdella grossa* (Jackson, 1935).

Subepidermal or subcutaneous sense organs, restricted entirely to the polystyliferous hoplonemerteans, consist of pear-shaped groups of cells arranged around a central lumen (Fig. 12B). From the proximal parts of the organ a single nerve extends into the body, leading to the dorsal ganglia via cephalic nerves. Coe (1926, 1927a) suggests that these may either represent degenerate eyes, although of quite different appearance, or be specialised sensory organs in some way related to the pelagic mode of life.

In nemerteans special sensory structures are limited principally to the anterior regions of the body. Lateral or oblique cephalic grooves, particularly obvious in hoplonemerteans, are formed by a decrease in epidermal height and may consist either of continuous furrows or rows of pits, lined with slender sensory ciliated cells. The cell cytoplasm is more dense than that of the normal epidermal cells. There are either one or two pairs; in the latter case the anterior pair are always in front of the cerebral ganglia. Conversely, cephalic slits, characteristic of the heteronemerteans, are deep lateral or transverse grooves lined by a modified epithelium similar to that of hoplonemerteans. Neither grooves nor slits contain gland cells or pigment granules. The epithelium of cephalic slits and grooves lies over a zone of ganglia cells, and it is usually assumed that both possess chemotactic functions. In the Lineidae the horizontal furrows are strikingly developed (Figs. 2F, 12H).

Carinomella, *Tubulanus* and some other palaeonemerteans possess a pair of lateral transverse grooves just at the back of the head, lined with a slightly differentiated epithelium that is apparently more sensory in nature than the remainder of the epidermis. The palaeonemertean condition, however, is not generally as well developed as in other groups.

Much of the anterior region between the cerebral ganglia and the front tip is filled with the frontal or cephalic glands, clusters of large vacuolated cells containing a variety of irregularly disposed, often acidophilic, inclusions. The glands commonly open into a frontal sensory organ (supra-oral sense organ) by ducts passing between the epithelial cells of the latter, or in other species they open independently via a single or numerous tubules connecting directly with the epidermis (Fig. 12C). The frontal organ characteristically occurs in hoplonemerteans as a flask-shaped protrusible pit opening at the extreme anterior tip of the body. It is lined by a non-glandular epithelium, whose cells are distally extended into long, rather bristle-like, cilia. Three small structures, histologically similar to the hoplonemertean frontal organ and supposedly a variety of it, are found at the anterior end of some heteronemerteans (*Cerebratulus*, *Lineus*, *Micrura*). In other heteronemerteans only one such structure is present. The frontal organs are innervated by numerous fibres originating from the cerebral ganglia, and are presumed to possess a chemotactic function.

The cephalic glands are best developed in the hoplonemerteans, especially in the terrestrial *Geonemertes* (Pantin, 1969) and parasitic *Carcinonemertes epialti* (Coe, 1905a), where they occupy the bulk of the anterior region. In some nemerteans they extend posteriorly beyond the cerebral ganglia, projecting over the gut (Friedrich, 1955), whilst in others they are much more restricted, forming only a short dorsal cap of cells (Coe and Kunkel, 1903; Gibson and Young, 1971; Moore and Gibson, 1973). The secretions of the glands are mucoid; they are particularly active in terrestrial species, producing copious amounts of a thick mucus that often envelopes the whole animal.

Eyes are present in most of the monostyliferous hoplonemerteans and in many palaeo- and heteronemerteans. They are always restricted in their distribution to the anterior region, mostly to the area in front of the cerebral ganglia (Figs. 2A,E–K). The number of eyes depends upon the species and ranges from two, to more than two hundred. They are arranged in median or paired clusters or rows, the precise number often varying between different individuals of the same species as well as with age. Adult *Cephalothrix linearis*, for example, are eyeless, but the young have two well-marked eye spots that disappear with growth. *Geonemertes australiensis* was reported as possessing thirty to forty eyes by Dendy (1892), but the examination of more than 500 specimens showed that the number could attain a value of 170 (Hickman, 1963). Some species, as in *Emplectonema bocki* (Brunberg, 1959), show an eye multiplicity, essentially the splitting of the basic eye number into several separate eyes rather than the development of new and completely independent structures. In the genus *Zygonemertes* the eyes are arranged in a loose row on either side of the body and extend for some distance behind the cerebral ganglia (Coe, 1905a).

The eyes are nearly always situated beneath the epidermis, in the dermis, musculature or parenchyma or, less commonly, directly attached to the cerebral ganglia. Their general structure is the same as found in turbellarians, namely that of an inverted pigmented cup ocellus. Each is formed of a single-layered epithelium, curved into a cup-shape, in which the cytoplasm contains black, brown or reddish pigment granules. The interior of the cup contains retinal or photo-receptive cells that are elongate structures, terminating in a rod border that is in direct contact with the pigmented cells. At the opposite end each retinal cell has a nucleated swelling, beyond which the eye nerve fibres pass out from the eye through the mouth of the cup, joining with fibres from other retinal cells to form an ocellar nerve originating from the cerebral ganglia. Vernet (1970) reports that in *Lineus ruber* each eye contains fifty photoreceptive cells, each provided with dendritic processes full of mitochondria, neuro-tubules, vacuoles and a solitary axial filament. In the terrestrial *Geonemertes agricola* the eyes are rather different, consisting instead of a closed ovoid body whose wall is formed from the pigmented epithelium. The nucleate portions of the retinal cells are aggregated outside this wall, their distal regions passing through the pigment zone into the fluid-filled interior. In these eyes the central fluid may possibly have a diffractive function. The most advanced nemertean eye structure is found in *Drepanophorus*, where each eye consists of a complete visual apparatus including lens, pigment cells, retina and optic nerve (Fig. 12D).

Statocysts are only found within a single hoplonemertean genus, *Ototyphlonemertes*. They consist of one or more pairs of ovoid vesicles, more spherical in swimming forms, lying in the posterior region of the brain on the dorsal surface of the ventral lobes (Figs. 12E,F). The vesicles are lined by a thin unciliated epithelium enclosing a spherical to dumb-bell shaped statolith or mass of small globules. In *O. pallida* subspecies *czerniavskyi* two pairs of statoliths occur in each statocyst. It appears possible that the statocysts are responsible for geotactic behaviour, comparable to that found in some acoele turbellarians (Corrêa, 1948, 1950).

Peculiar to and very characteristic of nemerteans are the cerebral organs (sometimes called cephalic organs), absent only from *Carcino-nemertes*, *Malacobdella*, the Polystylifera and some palaeonemer-teans. They consist of a pair of invaginated epidermal canals whose more or less spherical inner end is embedded in a mass of glandular and nervous material, closely associated with the dorsal or ventral cerebral lobes. They open to the exterior by way of the cephalic grooves or slits when these are present, or else via a simple pit on the body surface, positioned near the brain. Their simplest condition is found in some palaeonemerteans (*Carinina*, *Carinoma*, *Tubulanus*) (Fig. 12G), where they comprise simple lateral sensory pits (Thompson, 1900; Coe, 1905a). In other palaeonemerteans a short duct leads directly to a rather ovoid chamber (Bürger, 1895; Scharrer, 1941). Further development, found in other nemerteans, is achieved by elongation and curving of the duct (cerebral canal) to the extent that it penetrates the body-wall musculature into the parenchyma. Some palaeo- and the hoplonemerteans have an inner sensory portion of the cerebral canal embedded in a complex of closely packed nervous and glandular cells that is connected to the cerebral lobe by one or more large nerves (Fig. 12I). The most advanced development of the cerebral organs is found in heteronemerteans such as *Cerebratulus* and *Lineus*, where the glandular component fused with the dorsal lobe is so massive that it gives the impression of being an additional posterior ganglionic lobe (Scharrer, 1941; Brunberg, 1964; Ling, 1969a,b, 1970) (Fig. 12H). The cerebral organs often lie close to lacunar regions of the lateral blood vessels (Coe, 1895; Bürger, 1897–1907), but in *Hubrechtia* they penetrate into the vascular lumen a short way, whilst in some cerebratulids the blood sinuses almost completely surround the organs (Bürger, 1897–1907; Coe, 1905a). Gland cells in the more complex types may be arranged into definite groups or tracts. In *Drepanophorus* and its relatives the cerebral canal forks, one branch terminating in a blind-ending sac, the other continuing as a long curved cerebral duct.

Ling (1969a,b, 1970) found that in *Lineus ruber* the cerebral canal forms three right-angled bends within the organ lobule to terminate blindly in vesicular tissue composed of very large vacuolated cells possessing poorly defined boundaries (Fig. 12I). The canal is horizontally divided into major and minor canals by a ciliated septum, formed by the cohesion of dilated cilia protruding from special cells lining the canal walls. The cilia in the two canals beat in opposite directions, inwards in the major canal and outwards in the minor. At the first and second bends two groups of gland cells discharge their secretions. The whole structure is supplied by at least two types of bipolar nerve fibres joining the cerebral organ to a main nerve tract. Processes of some of the bipolar cells bear a single flagellum that penetrates into the canal lumen. The nerve tract is in turn connected via nerve fibres to ganglionated cells which complete the nervous junction with the brain.

The precise function of the cerebral organs is still not understood. They have variously been suggested as auditory structures (Quatrefages, 1846; Jensen, 1963), respiratory organs (Hubrecht, 1875, 1880b), excretory aids (McIntosh, 1876), chemotactic organs used in either the detection of food (Reisinger, 1926) or the analysis of water (Bürger, 1895), or as involved in some endocrine function (Scharrer,

c

1941). Water currents are certainly maintained within the canals, and there is some evidence that the ciliary rate increases in the presence of a suitable stimulus. Dewoletzky (1887) reported that the organs' secretions from *Cerebratulus* were very viscous and refringent. Willmer (1970) suggests that the cerebral organs form potential precursors for either the vertebrate eye or a component of the pituitary. Ultrastructural evidence is no more conclusive, as the anatomy of the organ is in keeping with a chemoreceptive, mechanoreceptive or photoreceptive function (Ling, 1969b).

In one nemertean species only (*Emplectonema kandai*) glandular luminescent organs have been reported within the epidermis (Kato, 1939).

Alimentary canal

The alimentary canal in all nemerteans consists essentially of a ciliated tube extending the full body length with the exception of a short cephalic region. The mouth, which is small in hoplonemerteans but large in most others, is positioned ventrally near the anterior tip, either in front of or immediately behind the cerebral ganglia, except in the palaeonemertean family Cephalothricidae, where it is much more posteriorly placed at the end of a long snout (Figs. 1A, 6A). The mouth and proboscis openings are separate in most pelagic hoplonemerteans, but in *Protopelagonemertes* and *Planktonemertes* both open on the subterminal portion of the head via a short atrium (Fig. 13A). At the posterior end of the atrium the dorsal opening leads to the rhynchodaeum, the ventral to the gut.

In the simplest forms, such as the palaeonemerteans *Hubrechtella*

Fig. 13
Alimentary and blood vascular systems. **A**, Vertical longitudinal section through anterior end of *Planktonemertes agassizii*; **B**, Anterior alimentary system of *Amphiporus occidentalis*; C–H, Blood vascular systems of *Cephalothrix linearis* (**C**), *Tubulanus annulatus* (**D**), *Carinoma armandi* (**E**), *Lineus sanguineus* (**F**), *Cerebratulus lacteus* (**G**) and *Amphiporus lactifloreus* (**H**). **at**, atrium; **cac**, cardiac caecum; **cl**, cephalic lacuna; **cvl**, cephalic vascular loop; **i**, intestine; **ic**, intestinal caecum; **icd**, intestinal caecal diverticulum; **lbv**, lateral blood vessel; **lc**, lacunar commissure; **lrv**, lateral rhynchocoel vessel; **lvb**, lateral blood vessel branches; **mbv**, middorsal blood vessel; **oca**, oesophageal caecal appendix; **oe**, oesophagus; **oec**, oesophageal caecum; **p**, proboscis; **pir**, proboscis insertion ring; **pt**, pyloric tube; **r**, rhynchocoel; **rbv**, rhynchocoel blood vessel; **rd**, rhynchodaeum; **sm**, stomach; **tvc**, transverse vascular connective. (A redrawn from Coe, 1926; B redrawn from Coe, 1905a; C redrawn from Bürger, 1897–1907; D–F, H redrawn from Oudemans, 1885; G redrawn from Coe, 1895)

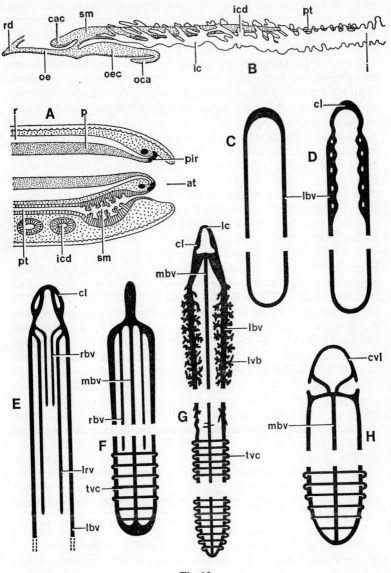

Fig. 13

and *Tubulanus*, the gut is unspecialised throughout its length, but in most it is both morphologically and functionally divisible into foregut and intestine (Kirsteuer, 1967a; Jennings and Gibson, 1969). The foregut is often further divisible into a buccal cavity, which frequently has a ventral epithelium several times thicker than the dorsal, an oesophagus, and a stomach. In palaeo- and heteronemerteans these divisions are not usually evident externally, but in hoplonemerteans a slender or broad oesophagus leads from either the rhynchodaeum or buccal cavity to an enlarged, often bulbous, stomach that commonly has deeply folded walls and is visible, at least in paler coloured species, through the body wall. In the polystyliferous hoplonemerteans the foregut is very short, with the oesophagus correspondingly reduced or even absent. Behind the stomach the gut either leads directly into the intestine, as in anoplan forms, or enters a long or short pyloric tube that connects it with the dorsal wall of the intestine, as in most hoplonemerteans (Friedrich, 1956). In the latter case the anterior end of the intestine extends forwards and ventrally to orm a blind-ending caecum. In *Paranemertes californica* the typical long slender pylorus of hoplonemerteans has disappeared, and a shortened pyloric tube opens into the anterior end of the long intestinal caecum. This is demonstrated by the fact that posterior to the caecum the intestine possesses a region with the same histological structure as the pylorus (Coe, 1905a). The gut of bdellonemerteans (*Malacobdella*) is quite atypical of the phylum (Riepen, 1933; Gibson and Jennings, 1969); in these species the rhynchodaeum is lost during development and the proboscis opens into the dorsal wall of the foregut. The anterior aperture thus is functionally like that of monostyliferous hoplonemerteans but differs in its development and anatomy. The foregut, or pharynx (Coe, 1905a, calls it the oesophagus), is a large barrel-shaped structure occupying nearly half of the body length. It is lined by longitudinal tracts of ciliated motile papillae that are capable of being interlocked during feeding (see Chapter 2). At its posterior end the pharynx narrows to a short oesophageal constriction lacking papillae that leads directly into the sinuous intestine. The hindgut does not possess lateral diverticula nor other pouches.

Most nemertean species have intestinal diverticula. They may be simple lateral protuberances, extending for only a short distance, or may branch distally and be quite long. In general they become progressively smaller as the anus is approached, and may be totally absent from the extreme anterior part of the intestine. There is usually a short, occasionally longer, region immediately in front of the anus that also lacks diverticula. The caecum in hoplonemerteans normally possesses diverticula similar in appearance to those of the intestine (Fig. 13B).

A few hoplonemertean species show an anatomical complication of the gut in the development of additional pouches. *Emplectonema intestinalis* has a dorsal appendix to the stomach that extends below the ventral rhynchocoel floor (Friedrich, 1958) and a few *Amphiporus* (*Intestinonemertes*) species have a distinct oesophageal caecum projecting ventrally and posteriorly from a point just behind the ventral cerebral commissure (Coe, 1905a; Friedrich, 1957). In *Amphiporus occidentalis* the stomach additionally extends forwards above the oesophagus to form a broad, bulbous chamber, the cardiac caecum (Fig. 13B).

The epithelium of the lips and buccal cavity resembles the epidermis in consisting of attenuated epithelial cells with filamentous bases, interspersed with gland cells of two basic types. Amongst anoplan forms the gland cells produce and secrete either mucus or acidic substances, but in the Enopla a physiological modification has occurred and, although histologically like the gland cells of palaeo- and heteronemerteans, somewhat different secretions are formed. In some nemerteans (*Cerebratulus*, *Micrura*) the beginning of the buccal cavity is encircled by a ring of long-necked salivary or accessory buccal glands. These are placed in the outer longitudinal musculature but open into the buccal cavity (Coe, 1901). The oesophagus, when present, has a much thinner epithelium lacking gland cells, but the stomach is richly glandular and either duplicates the buccal secretions, as in palaeo- and heteronemerteans, or else possesses its own range of glandular products, as in the other orders. The gland cells in general fall into one of two categories, either being basophilic, with coarse to fine granular or homogeneous contents which are not enzymic, or acidophilic and filled with numerous small spherical globules. In those nemerteans in which stomach enzymes have been reported they have been confined to these acidophilic glands (Jennings, 1962; Jennings and Gibson, 1969). In the Bdellonemertea a somewhat different situation is found, with many gland cells being migratory and subepithelial, discharging their contents to the pharyngeal lumen via tracts extending between epithelial cells (Gibson and Jennings, 1969). In all species the foregut is densely ciliated, this being one of the characters that enables the two major gut regions to be distinguished.

The intestinal epithelium or gastrodermis, considerably thicker than the foregut in most nemerteans, is composed of attenuated ciliated cells containing a variety of inclusions, interspersed with pyriform acidophilic gland cells. The inclusions are mostly fat or oil globules, or vacuoles in various stages of digestion, but some species possess crystalline structures and in at least one (*Paranemertes californica*) there are dark green pigment granules (Coe, 1905a). The anterior intestine of *Hinumanemertes* does not contain any gland cells (Iwata, 1970).

The intestinal cilia are long, sparsely distributed, and contain between their proximal regions typical dense microvilli (Jennings, 1969). In hoplonemerteans, besides the usual gland cells, there are found acidophilic vesicles situated in the cytoplasm of the ciliated cells. These vesicles, which are secretory, are produced proximally but migrate distally to discharge their contents at the appropriate phase of digestion, then becoming absorbed by the cytoplasm (Jennings and Gibson, 1969; Gibson, 1970).

The number of intestinal gland cells varies depending both upon the species and region concerned. Commonly they are most abundant towards the front of the intestine, decreasing in density posteriorly. In some species gland cells are absent from the region immediately in front of the anus, this portion then being termed the rectum. *Malacobdella* has very abundant gland cells distributed throughout the intestinal length, extending right up to the slightly protuberant anus.

The gut does not usually possess its own musculature, but in several species the body-wall musculature lies close to or immediately against the gut epithelium. Where the inner body-wall muscle layer is composed of circular fibres, as in palaeonemerteans, this embraces both the gut and rhynchocoel in a single muscular sheath. In other forms either the inner longitudinal muscle fibres lie against the gut, or the entire body-wall region is separated from it by thick or thin parenchyma. The foregut of *Hinumanemertes* has thin inner circular and outer longitudinal muscles associated with its ventral and lateral walls (Iwata, 1970).

Some muscle fibres, mainly of the longitudinal type, are found in close association with the posterior region of the foregut or stomach. These are derived from, but are distant to, the body-wall musculature. In some species an extremely thin development of circular fibres is found in conjunction with part or most of the intestine. Coe (1926) reports that some pelagic hoplonemerteans have special muscle bands inserted into the stomach walls; these apparently serve to hold the gut in place within the gelatinous parenchyma during proboscis eversion, and may also assist mechanically during ingestion.

Blood system

The blood system of nemerteans is of the closed type and consists of two types of vessels, the blood lacunae that are spacious channels lined only by a thin membrane, and the vessels proper, which have definite walls composed of two or more layers. Both types of vessels are embedded in the general parenchyma.

Various grades of complexity in the development of the blood system in nemerteans can broadly be correlated with the taxonomy of the phylum. The simplest condition is found in the palaeonemertean family Cephalothricidae (Oudemans, 1885), where the blood system comprises a pair of lateral longitudinal vessels, running slightly beneath the gut, joined anteriorly and posteriorly by a cephalic and anal lacuna respectively. There are no major branches at this level (Fig. 13c).

In other palaeonemertean families various additions to the basic form are produced principally by the development and subdivision of the cephalic lacuna (*Carinesta*, *Carinomella*), by the formation of a subdivided oesophageal lacuna in the region of the foregut (*Tubulanus annulatus*) (Fig. 13d), and by the introduction of a pair of rhynchocoel vessels that originate from the lateral vessels by multiple branching and extend posteriorly as ridges in the ventrolateral rhynchocoel walls. An additional pair of lateral vessels, more dorsally positioned, are found in the oesophageal region in *Carinoma* (Fig. 13e) but these do not have branches extending into the rhynchocoel (Bergendal, 1903). In the more complex forms transverse pseudometameric junctions occur between the lateral blood vessels, uniting these above the intestine (Fig. 13f). The family Hubrechtidae is the only palaeonemertean group to possess a dorsal blood vessel.

In most other anoplan nemerteans there are three basic longitudinal vessels (paired lateral and single middorsal) that branch more or less freely. All three are connected anteriorly by an anastomosing lacuna, which may be enormously spacious as in *Hinumanemertes* (Iwata, 1970), and in many species are further joined throughout the intestinal region by more or less regular transverse connections situated above and immediately outside the intestinal diverticula. In some genera (*Cerebratulus*) there are additional vessels and lacunae in the oesophageal region (Coe, 1895) (Fig. 13g). At the posterior end the middorsal vessel runs into the anal lacuna. Rhynchocoel and dorsolateral vessels are not usually found in heteronemerteans, although they do occur in some species.

A complex anterior blood system is found in the heteronemertean *Hinumanemertes kikuchii* (Iwata, 1970). The large cephalic lacuna begins near the anterior tip as a compressed crevice over the rhynchodaeum. It extends posteriorly and ventrally beside the rhynchodaeum, enclosed by the rhynchodaeal longitudinal musculature. In the middorsal region of the lacuna, bundles of dorsoventral muscle fibres divide the lacuna into three or four distinct chambers, but near the cerebral septum the dorsal chambers disappear and a pair of cephalic lacunae extend posteriorly towards the brain. These pass through the cerebral septum as two narrow canals, which fuse into a flattened cerebral lacuna lying beneath the anterior rhynchocoel. A middorsal blood vessel originates from this lacuna to run inside the proboscis sheath up to the anterior intestine. The cerebral lacuna is divided into three chambers by transverse muscle bundles running below the proboscis sheath; two chambers are placed laterally, one ventrally. The former extend posteriorly and do not lose their lacunar form until reaching the intestinal region, where they narrow into the lateral blood vessels that are transversely connected with the mid-dorsal vessel.

The basic plan of three longitudinal vessels is also found in the hoplonemerteans, together with the transverse intestinal junctions (Fig. 13H). Valves, simple flap-like extensions incompletely traversing the vascular lumen, are a characteristic feature of hoplonemertean blood vessels, but are absent from those species with more lacunar-like systems. Some simplification of the blood system occurs in most hoplonemerteans by the cephalic and foregut lacunar network disappearing to be replaced by simple cephalic vessels that form an anterior loop (Fig. 14A). The fundamental vascular pattern is retained by the pelagic families except for the Pelagonemertidae, where the dorsal vessel is limited to a short rudimentary anterior branch of the cephalic loop (Coe, 1926). The longitudinal vessels are joined by simple anastomoses, anteriorly by one dorsal and one ventral junction (except in *Proarmaueria pellucida*, where the dorsal anastomosis

Fig. 14

Blood vascular system. A–E, Blood vascular systems of *Prostoma graecense* (A), *Nectonemertes pelagica* (B), *Planktonemertes agassizii* (C), *Proarmaueria pellucida* (D) and *Pelagonemertes brinkmanni* (E); F, Vascular plugs of *Geonemertes dendyi*; G, Transverse section through typical nemertean contractile blood vessel. bmb, part of main blood vessel beneath vascular plug; cbv, caudal blood vessel; cvl, cephalic vascular loop; eda, evagination of dorsal cephalic anastomosis; end, endothelium; gl, gelatinous layer; lbv, lateral blood vessel; mbv, mid-dorsal blood vessel; ml, muscular layer; oal, outer anucleate covering layer; py, parenchyma; ren, rhynchocoel endothelium; vca, ventral cephalic anastomosis; vp, vascular plug; x, loop in middorsal vessel marks point where vessel leaves rhynchocoel wall. (A after Böhmig, 1898; B–E redrawn from Coe, 1926; F redrawn from Pantin, 1969; G redrawn from Gibson and Jennings, 1967)

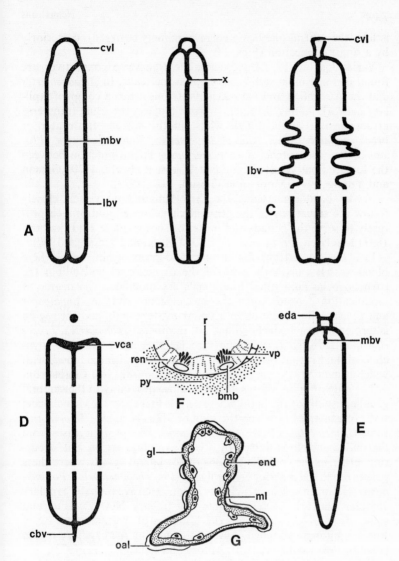

Fig. 14

is lost, the ventral one being correspondingly enlarged), posteriorly by a single connection (Figs. 14B–E).

Variations on the pseudometameric transverse connectives are found in some monostyliferous hoplonemerteans. In *Acteonemertes* and the Australian and New Zealand *Geonemertes* a complex capillary network joining the lateral vessels replaces the usual transverse arrangement (Pantin, 1969), whilst in the various freshwater or brackish-water forms (*Campbellonemertes*, *Potamonemertes*, *Prostoma*, *Sacconemertella*, *Sacconemertopsis*) lateral junction between the longitudinal vessels is completely absent (Iwata, 1970; Gibson and Young, 1971; Moore and Gibson, 1972, 1973).

In the bdellonemerteans the three longitudinal vessels closely follow the sinuations of the gut and rhynchocoel, and branch profusely to serve the gonads and posterior sucker (Plate 2) (Takakura, 1897; Maclaren, 1901; Riepen, 1933; Gibson and Jennings, 1967).

In a few nemerteans for certain, and probably in most, where blood vessels run in the wall of the rhynchocoel and adjoin the rhynchocoelic fluid, their lining cells are modified. The degree of modification depends upon the species concerned; in *Amphiporus* and *Lineus* cellular alteration is comparatively slight and the region is termed a rhynchocoelic villus, but in others (*Geonemertes*, *Prosorhochmus*, *Prostoma*) a definite vascular plug is formed. The rhynchocoel fluid is thus brought into close contact with the blood, with which it may have some physiological relationship (see Chapter 3).

The plug or villus always occurs in close proximity to the cerebral ganglia, usually at the point where vessels first enter the rhynchocoel wall. In examples in which the plug or villus is single (*Amphiporus*, some *Geonemertes*, *Lineus*, *Prosorhochmus*, *Prostoma*), it arises from the middorsal blood vessel. In others two plugs are present, occurring either in the cephalic regions of the lateral vessels where these pass through the anterior cerebral ring (*Campbellonemertes*, *Potamonemertes*), or, in species which possess an extensive cephalic capillary vascular network (*Acteonemertes*, Australian, New Zealand and some other *Geonemertes*), on the branches of the middorsal vessel where it bifurcates immediately before entering the capillary region (Pantin, 1969; Willmer, 1970; Moore and Gibson, 1972, 1973).

The larger blood vessels have four layers to their walls (Fig. 14G). These are a lining endothelium of flat or slightly ovoid cells, a thick connective tissue layer of homogeneous appearance, a layer of circular muscle fibres, and an external covering layer which lacks nuclei. These vessels are contractile, differing from the non-contractile types in which the endo- and epithelium are separated simply by a thin membrane without any muscle fibres. The valves of hoplonemertean blood vessels are formed from cup-shaped cells contained in the gelatinous connective tissue. When the vessel contracts these cells project into the lumen, but their exact role in the circulation of the blood is not understood.

Nemertean blood is mostly colourless but may be yellow, green, red or orange. It consists of a clear fluid in which the blood corpuscles float, these being small oval or rounded nucleate discs that are capable of some slight alteration in shape. Frequently amoeboid lymphocytes, similar in appearance to the parenchymal wandering cells, are also present. Ohuye (1942) recognises four types of lymphocytes in *Lineus* (Fig. 20).

In those nemerteans with coloured blood the colour occurs in the corpuscles. Red pigment, according to Lankester (1872), may be haemoglobin, a respiratory pigment that in invertebrates is commonly associated with anaerobic environments (Hyman, 1951), but several species without coloured blood possess similar pigments around the cerebral ganglia and along the margins of the lateral nerves.

Excretory system

A nephridial excretory system is present in most nemerteans, although none has been found in the pelagic hoplonemerteans (Brinkmann, 1917; Coe and Ball, 1920; Coe, 1926, 1927a, 1930a). In their simplest development they consist of a single pair of long branched ducts, generally limited to the anterior region, opening on the body surface by a single nephridiopore on either side. The duct extends anteriorly from the pore, and is sometimes found in close association with the lateral blood vessels or anterior lacuna, into which it may partially project. Most palaeonemerteans have a broad undivided nephridial duct confined to the posterior half of the oesophageal region (*Carinoma, Carinomella, Tubulanus*), that has numerous flame cells deflecting the cephalic blood-vessel wall and comprising a nephridial gland (Coe, 1905a) (Figs. 15A,B). In some species the vascular wall is absent so that the flame cells are directly bathed by the blood, although they do not open into it (Coe, 1943). The cephalothricid palaeonemerteans *Cephalothrix major* and the females of *C. spiralis* have a far more extensive excretory system developed, and the single longitudinal duct on either side is replaced by a multiplication of the nephridia, each of which possesses its own efferent tubule to the body surface. More than 300 pairs of nephridia have been found in a single animal slightly longer than one metre (Coe, 1930a,b,c). Each nephridium (Fig. 15c) consists of a multinucleate flame cell (incorrectly called a metanephridium by Coe), possessing a mushroom-shaped head (nephrostome) and thin-walled duct that leads to a narrow neck

67
6 Nemerteans

joining the broad convoluted tubule, that in turn is connected to its own nephridiopore by a slender radial efferent duct. The convoluted tubule is apparently directly comparable with the usual longitudinal duct of other forms. The flame cells are placed close to the lateral blood lacuna. The cytoplasm of the convoluted tubule walls is coarsely granular and contains numerous inclusions; it is syncytial, the nuclei not being separated by distinct cell boundaries. The tubule is lined by long slender cilia.

In contrast, male *Cephalothrix spiralis* have the more conventional protonephridial system with uninucleate and much simpler flame cells hanging freely in the blood vessels. This sexual dimorphism in the excretory system has not been recorded for any other nemertean species.

In *Carinella* (*Tubulanus*) *albocincta* there are five to eight longitudinal ducts on either side, joined together only near the efferent duct where they form a large lacuna. Where flame cells penetrate or distort the vascular lining, they are often positioned on low papilla formed from the blood-vessel endothelium.

Most other palaeonemerteans, and the hetero- and hoplonemerteans, usually possess nephridia that are prorfusely banched, with numerous capillaries and flame cells closely applied to the lateral vessels or foregut lacunae (Fig. 15D). In the *Geonemertes pelaensis* group the dense aggregation of flame cells around the cephalic blood vessels is analogous with the nephridial gland of palaeonemerteans. The entire excretory system may be confined to the cephalic or anterior region or, as in some species (*Geonemertes, Prostoma,*

Fig. 15
Excretory system. **A**, Nephridial system of *Carinoma mutabilis*; **B**, Transverse section of lateral blood vessel and nephridial gland of *Carinoma mutabilis*; **C**, Nephridium of *Cephalothrix major*; **D**, Typical nemertean nephridial arrangement; **E**, Nephridial duct system of *Prostoma graecense*; **F**, Nephridium of *Geonemertes agricola*; **G**, Flame cell of *Geonemertes rodericana*; **H**, Flame cell of *Geonemertes dendyi*. **bvw**, blood-vessel wall; **cfl**, ciliary flame; **con**, convoluted tubule; **csp**, cuticular support; **ctu**, collecting tubule; **ecl**, end canal; **ed**, efferent duct; **fc**, flame cell; **fcn**, flame-cell nucleus; **lb**, longitudinal bar; **lbv**, lateral blood vessel; **lnc**, large nephridial canal; **mbv**, middorsal blood vessel; **mlc**, main longitudinal canal; **ngt**, nephridial gland tubule; **nnc**, narrow nephridial canal; **np**, nephridiopore; **ntc**, nucleus of terminal chamber; **tb**, terminal bulb; **td**, terminal duct; **tr**, transverse ring. The arrows in A indicate the level of the junction between stomach and intestine. (**A, B** redrawn from Coe, 1905a; **C, D** redrawn from Coe, 1930a; **E** after Böhmig, 1898; **F** basedon Coe, 1930a and Pantin, 1969; **G, H** redrawn from Pantin, 1969)

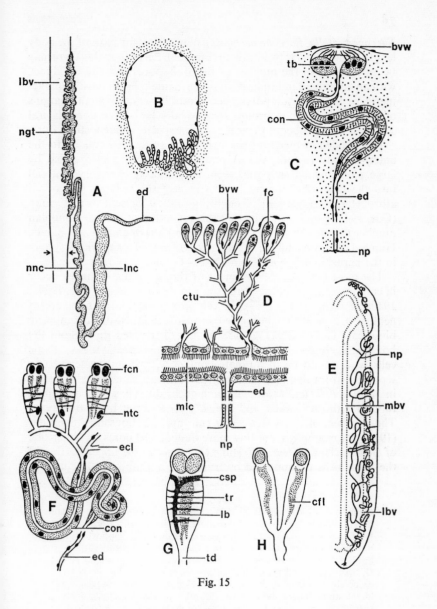

Fig. 15

Sacconemertella, Sacconemertopsis), may extend for most of the body length (Pantin, 1969; Iwata, 1970) (Fig. 15E). Paralleling this there may be an increase in the number of nephridiopores and efferent ducts, with an additional multiplication of the flame cells. In most hoplonemerteans the nephridiopores open ventrally near the lateral margins of the body, in other groups they usually lie on the dorso-lateral surfaces. The number of pores is for some species a diagnostic feature.

In the bdellonemerteans the nephridial system is confined to the anterior body regions, the excretory tubules being surrounded by large parenchymal cells that seem to discharge excretory material into the nephridial lumen. In this respect the cells are functional athrocytes. A single pair of nephridia only is present in this group (Coe, 1945b), and it may be that this reduction of the usual enoplan number can be related to the genus' adoption of commensal habits. Humes (1942) was unable to find any evidence of a nephridial system in the parasitic hoplonemertean genus *Carcinonemertes*.

The most complex development of the nephridia seems to occur in certain examples of the terrestrial genus *Geonemertes* (Coe, 1929a, 1940a; Pantin, 1947, 1969). In this group very large numbers of nephridia, typically containing either paired uninucleate or composite binucleate flame cells (Figs. 15F–H), are distributed throughout the body parenchyma. With the exception of the cephalic region, flame cells tend not to be in intimate contact with blood vessels, being separated from them by parenchymal tissue, a situation also found in *Malacobdella* and *Prostoma*. Each nephridium has two convoluted tubules with thick walls, and opens to the exterior by its own pore. The numbers of pores frequently run into the thousands, Schröder (1913), for example, recording more than 35,000 pores on either side of *Geonemertes pelaensis*. In some *Geonemertes* the upper portion of the flame cells is supported by transverse and apparently cuticular

---

Fig. 16

Gonads. **A**, Gonad distribution in *Tetrastemma cincum*; **B, C**, Gonad distribution in male (**B**) and female (**C**) *Pelagonemertes brinkmanni*; **D**, Testes and ovaries of *Coenemertes caravela*; **E**, Optical section through spermary of *Nectonemertes mirabilis*; **F**, Ventral surface of head of *Phallonemertes murrayi* showing five pairs of papillary penes; **G**, Spermaries and penes of *Phallonemertes murrayi*. **bw**, body wall; **cg**, cerebral ganglia; **d**, dermis; **dsz**, developing spermatozoa; **g**, gonad; **id**, intestinal diverticulum; **m**, mouth; **ml**, muscular layer; **msz**, mature spermatozoa; **o**, opening from spermary into seminal vesicle; **ov**, ovary; **pl**, papillary penis; **r**, rhynchocoel; **sd**, spermatids; **sg**, spermatogonia; **sp**, spermary; **sv**, seminal vesicle; **sy**, primary spermatocytes; **t**, testis. (A modified from Corrêa, 1957; B, C modified from Coe, 1926; E–G redrawn from Coe, 1926; D redrawn from Corrêa, 1966)

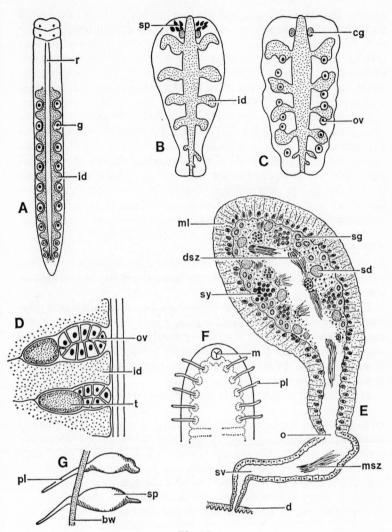

Fig. 16

skeletal bars (Figs. 15F,G), and the number of flame cells associated with each nephridium is increased to between six and ten.

A rather aberrant condition is found in some heteronemertean species (*Baseodiscus*) in that the single pair of nephridial ducts has several efferent tubules, some of which open to the exterior in the usual way whilst others open via an internal duct into the oesophageal lumen (Punnett, 1900; Coe, 1906).

Closely related species may show quite distinctive differences in the arrangement of their nephridial systems. For example, the tubules of *Cerebratulus lacteus* are profusely branched and closely linked to the lateral blood vessels of the foregut region, whereas those of *C. marginatus* are embedded in the walls of large blood lacunae. Both of these species possess a single efferent duct, whereas *C. melanops* has six or more. When duct duplication occurs, the number often increases with body size (Coe, 1943).

Reproductive system

Most nemerteans possess separate sexes but a few are hermaphroditic, particularly those hoplonemerteans with terrestrial or freshwater habits.

The gonads are spherical to flask-shaped sacs, usually and primitively confined to the intestinal region, where they form two lateral rows in the parenchyma alternating singly or in groups with the intestinal diverticula (Fig. 16A). In those palaeonemerteans without diverticula (*Tubulanus*) the gonads remain more or less serially repeated.

The adoption of parasitic or commensal habits has resulted in an increase in the gonad number in *Carcinonemertes*, *Gononemertes*, *Nemertopsis actinophila* and *Malacobdella*, with a concurrent loss of regularity so that they tend to become scattered throughout the general parenchyma (Bergendal, 1900; Coe, 1902a,b). A secondary reduction in the number of gonads has occurred in *Malacobdella minuta* (Coe, 1945b). In *Carcinonemertes* the gonads extend forward nearly as far as the cerebral ganglia (Coe, 1902b).

Repositioning of the gonads is found particularly amongst the pelagic species (*Paradinonemertes*, *Planonemertes*), where testes are limited to the cephalic or foregut region (Fig. 16B), beside or immediately behind the brain. The ovaries remain in the more primitive position between the intestinal diverticula (Fig. 16C), the testicular migration being regarded as a morphological advancement.

A rather atypical condition is found in the freshwater hoplonemertean *Potamonemertes* in that the testes are restricted to lateral rows of five to eight per side in the vicinity of the foregut, usually just dorsal to the lateral nerve cords. The ovaries are very numerous and extend for much of the body length; they overlap the testes distribution anteriorly but are in general confined to the interdiverticular spaces (Moore and Gibson, 1973). A similar situation occurs in *Dichonemertes* (Coe, 1938).

The gonads of *Coenemertes* (Corrêa, 1966) occur as paired double rows (Fig. 16D) with a tendency to show regional localisation of gamete forms, since more of the anterior gonads are functional testes than the posterior.

Several nemerteans with separate ovaries and testes are apparently variable or true protandric hermaphrodites, depending upon the environmental circumstances. Other hermaphroditic species have both male and female products formed within a common gonad, but these species frequently possess a strikingly protandric behaviour and may be developing towards separate sexes.

Sexual dimorphism occurs in some nemerteans, most commonly with respect to colour. This is clearly seen in *Malacobdella grossa*, where ripe males possess pinkish-cream testes, in contrast to the olive-green ovaries of gravid females. Immature specimens cannot similarly be differentiated on the basis of colour (Gibson, 1968). Morphological differences occur in *Nectonemertes* where the males have a pair of anterior tentacles developing at the onset of sexual maturity; similar structures are never found in females (Cravens and Heath, 1906; Coe, 1920). The bathypelagic species *Plotonemertes adhaerens* and *P. aurantiaca* have a pair of large posterior musculo-glandular appendages on the ventral body surface that may serve as gripping organs during mating (Brinkmann, 1917; Coe, 1936, 1945a). Sex differences in certain cutaneous gland cells are reported for *Lineus ruber* by Bierne (1970a).

The gonads originate from parenchymal cells that become clumped together before developing into a thin-walled sac filled with sex cells. The outer surface of the gonad may be lined by muscle fibres. Within the ovaries some cells may be involved in the production of yolk for the nutrition of the developing embryo. There is usually one gonad in each interdiverticular space, but in some species several ovaries or testes occur at each site, as in *Carcinonemertes*. In small forms producing comparatively large eggs (*Prostoma*, *Tetrastemma*) several intestinal diverticula may separate each ovary.

At the onset of sexual maturity a simple short duct grows out from the gonads to open at the body surface. The genital pores form a lateral or dorsolateral row down each side of the body; only rarely do they open ventrally other than in those nemerteans with scattered gonads (the commensal and parasitic species), in which the genital pores are irregularly distributed over the appropriate surface.

In *Nectonemertes* and *Pelagonemertes*, and to a lesser extent in some other pelagic genera, the testes (or spermaries) are ensheathed by a thick zone of special dorsoventral musculature (Coe, 1926) (Fig. 16E). This musculature is used in ejecting ripe spermatozoa during mating, the gonad being distally formed into a distinct tube which may have a small sperm reservoir near its terminal pore. A simple sort of copulatory organ is found in the genus *Phallonemertes* (=*Bathynectes*) (Brinkmann, 1912; Coe, 1926, 1936, 1945a). The sperm ducts are extended into slender muscular penes that project beyond the epidermis (Figs. 16F,G), forming five to seven pairs ventrally on either side of the head. In male *Carcinonemertes* the sperm ducts from the various testes enter a median longitudinal canal (Takakura's duct) that discharges into the posterior end of the intestine. Sperm is thus emitted via the anus (Humes, 1941, 1942).

In the anoplan nemerteans several (up to fifty) eggs are usually produced simultaneously within each ovary, whereas hoplonemerteans tend to ripen one egg at a time and lay them at intervals. The eggs are released from the ovaries by muscular contractions of the body walls squeezing the gonad, but in some instances release is accomplished by the entire body rupturing.

The stimulus to breed is apparently a chemotactic one, males and females either spawning without physical contact, or else males crawl over females whilst discharging their ripe sperm. In several species two or more gravid individuals enclose themselves inside a common mucoid sheath, into which ova and sperm are shed. Fertilisation may be external or internal. Self-fertilisation is believed to be possible within some of the true hermaphrodites.

Ovoviviparity is found in some species of *Lineus*, *Poikilonemertes* and *Prosorhochmus* and in *Geonemertes agricola*, the eggs developing within the parent to give rise to miniature worms (Coe, 1904; Stiasny-Wijnhoff, 1942; Friedrich, 1955). In most cases the eggs are laid in mucoid or gelatinous epidermal secretions deposited as strings or irregular masses. The eggs may be separate or grouped into semispherical capsules originating from the ovarian wall. A definite breeding season is normally found, but this varies according to the species. In the northern hemisphere late spring to early summer is a common breeding time. The commensal *Malacobdella* breeds throughout the year, but has two reproductive peaks of activity that can be correlated with the biannual blooms of its phytoplankton food (Gibson, 1968).

Descriptive references

Many useful papers have been published that give descriptions of

numerous nemertean genera and species. Amongst the more recent
those of value to nemertean taxonomists include Brunberg (1964),
Corrêa (1954, 1956, 1957, 1958, 1963, 1964), Friedrich (1955, 1958,
1960, 1970), Iwata (1952, 1954a, 1970), Kirsteuer (1963a, 1965,
1967a) and Pantin (1969).

2

NUTRITION AND DIGESTION

Diet

Nemertean worms are for the most part carnivorous or scavenging macrophages, many species possessing extremely voracious appetites and active predatory habits. Natural foods reported for various species include a variety of protozoans, turbellarians, nematodes, annelids, crustaceans, molluscs and fishes, but under laboratory conditions some forms have been successfully maintained on diets that have included cooked beef fat, starch paste and homogenised liver (McIntosh, 1873–4; Coe, 1943; Corrêa, 1948; Gontcharoff, 1948; Hylbom, 1957; Jennings, 1960, 1962; Brunberg, 1964; Iwata, 1967; Gibson and Jennings, 1969; Jennings and Gibson, 1969; Gibson, 1970). Nemerteans do not normally appear to possess highly specialised diets, although few are as cosmopolitan in their choice of food as the heteronemertean *Lineus corrugatus*. Dearborn (1965) reports that under artificial conditions this species readily takes diatoms, sponges, anemones, polychaetes, amphipods, isopods, gastropods, fish, seal skin and blubber, and even sardines in tomato sauce. The vast majority seem to feed on a much less diverse range of organisms, and a few species show distinct dietary preferences. Du Plessis (1893) states that *Emea* (= *Prostoma*) is a nocturnal feeder, readily taking small and delicate crustaceans such as *Cyclops*, but showing a definite preference for the larvae of the lacustrian dipterans *Chironomus* and *Tanypus*. Similarly, Roe (1970) records that the hoplonemertean *Paranemertes peregrina*, whilst exhibiting a strong liking for the annelid *Platynereis bicanaliculata*, will also feed on a number of other polychaete species should the preferred form be unavailable. A few nemerteans, such as *Amphiporus lactifloreus* which under laboratory conditions will feed only upon the littoral amphipod

Gammarus locusta (Jennings and Gibson, 1969), appear far more restricted in their choice of food, and will succumb to starvation even when a plentiful supply of other associated animals is provided.

Atypical diets are found amongst nemerteans in a few instances. The ectoparasitic *Uchidana parasita* seems to feed on gill tissue of the bivalve host (Iwata, 1967), and the entocommensal *Malacobdella grossa* is an unselective microphagous omnivore, although occasionally able to supplement its diet by catching larger particles, especially crustacean larvae (Gibson and Jennings, 1969).

Observations on the nutrition of juvenile nemerteans have been confined principally to those species which have either a Desor larva (*Lineus*) or a pelagic pilidium larva (*Cerebratulus, Hubrechtella, Micrura*). Gontcharoff (1959) records that a liver diet permits a rapid growth rate in young *Lineus ruber* and *L. viridis*, but that an improved longevity is obtained when they are reared on mixed microfauna shaken from the algae *Cystosira* and *Pterocladia*. A culture of the diatom *Nitzchia closterum* proved completely unsuitable for these lineids, but Gontcharoff suggests that for *Lineus ruber* unhatched eggs would form an excellent diet in the early stages of their life. This idea is based upon an observation by Schmidt (1934) that whereas *Lineus viridis* eggs hatch directly from their gelatinous capsule, those of *L. ruber* do not and the unhatched eggs are used as food by the foraging young. Support for this suggestion comes from the fact that whilst young *L. ruber* require food immediately on leaving their egg cases, juvenile *L. viridis* do not feed for two to three weeks after hatching. The larvae of *L. albocinctus* and *L. bilineatus*, as well as of many other heteronemerteans, devour their larval envelopes after metamorphosis, although juvenile *Hubrechtella dubia* leave their egg cases without eating them (Cantell, 1966a, 1969).

An unusual case was reported by Joubin (1914) in which the hoplonemertean *Amphiporus incubator* lays its eggs inside a mucilaginous sheath, the female then degenerating under the histolytic influence of oesophageal phagocytes to form food for the freshly hatched young. These eventually free themselves from their larval envelope and become normal predatory carnivores.

Several authors, in studying the development of nemertean pilidium larvae, have observed that they feed on small particles by means of a ciliary feeding mechanism. Cantell (1969) describes this mechanism for a number of pilidium species in some detail. Particles coming under the influence of the apical plate cilia are carried along the sides of the helmet towards the margins of the lobes (Fig. 17). Strong ciliation, giving rise to a fast inwardly directed current, then carries the particles ventrally to the vicinity of the mouth. Here ciliary activity increases and food is carried rapidly into the buccal cavity. The larvae possess a sorting mechanism that employs the two major ciliary currents, and relatively large particles are guided away from the mouth. The strength of the currents can be assessed from the fact that even energetically moving flagellates are unable to escape once they have fallen under the influence of the ciliary activity. Food particles entering the mouth are carried by further ciliary action back through the gut to the junction between oesophagus and stomach (the intestinal valve of Coe, 1899a), when they are then 'swallowed' little by little and passed into the intestine by contractions of the oesophageal walls.

Detection of the food

Among adult nemerteans several species are capable of detecting their food at a distance, whereas others require the close proximity or even mechanical contact of the prey before responding by commencing feeding behaviour. Reisinger (1926) noted that the freshwater *Prostoma* could accurately locate food in still water at a range of 17 cm, presumably by means of some chemotaxis mediated via the cerebral organs, since the cilia of these structures show an accelerated

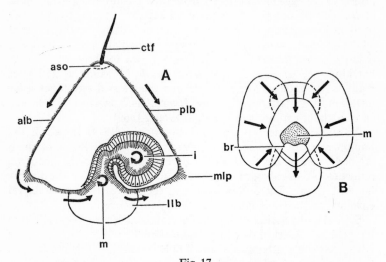

Fig. 17

Ciliary feeding currents of nemertean pilidia larvae. Arrows indicate routes taken by food particles. A, Lateral view; B, Oral view. **alb**, anterior lobe; **aso**, apical sensory organ; **br**, buccal ridge; **ctf**, ciliary tuft; **i**, intestine; **llb**, lateral lobe; **m**, mouth; **mlp**, marginal lappet; **plb**, posterior lobe. (A after Cantell, 1969; B redrawn from Cantell, 1969)

beat in the presence of food material. This inference is supported by Kipke (1932) who found that although decapitated *Prostoma* responded to the presence of both food and chemical substances, their reaction became much more orientated as the cerebral organs regenerated.

The condition of the food appears to be of some importance in determining the distance over which it can be detected. *Lineus ruber* and *L. sanguineus* apparently rely only upon visual contact for detecting intact living prey at a range of 2 to 3 cm, but are able to scent damaged animals or suitable organic material at a distance of up to 8 cm (Jennings and Gibson, 1969).

The possession of a chemotactic response does not necessarily indicate that the species can accurately track its food. Coe (1943) states that *Cerebratulus* only occasionally follows a direct line towards suitable food, even though stimulated by its presence to leave its burrow and commence foraging. Similarly, Beklemishev (1955) suggests that tubicolous polychaetes can be detected by nemerteans by means of their 'odour' passing out with the externally directed water currents, although they are not subsequently traced accurately to their burrows. The usual effect of chemotactic detection seems to be to stimulate the worms to move faster, at the same time as the head darts to and fro in exploratory movements, as seen in *Cephalothrix* species (Jennings and Gibson, 1969). Such a behavioural response necessarily increases the chance of a nemertean moving close enough to a potential prey to be able to catch it.

In the eyeless forms, such as many of the palaeo- and heteronemerteans, the food must presumably be located either chemically or by chance, although it is possible that some other taxis or kinesis, such as the detection of mechanical water disturbances, may be employed. Hyman (1951) observes that the frontal organs are everted in some species when food is close, although whether these are used in detecting prey prior to capture, or testing them for suitability afterwards, is not certain.

In contrast, neither the cerebral organs nor the eyes are used by other species that possess these structures, and food detection is apparently achieved purely by chance when the nemertean blunders into suitable organisms. This situation occurs in the hoplonemerteans *Paranemertes peregrina* (Roe, 1970) and *Amphiporus lactifloreus* (Jennings and Gibson, 1969), although for both of these species the environmental conditions at the time of feeding are such that the efficiency of food capture is not apparently impaired. *Paranemertes* feeds at low tide on the surface of mud-flats when the polychaete prey are incapable of escaping, and *Amphiporus* moves most rapidly in darker conditions when its amphipod food is relatively inactive and gregarious.

Coe (1943) observes that nearly all nemerteans actively search for their food, but this could be anticipated whether food contact is achieved intentionally, via some taxis, or accidentally. The microphagous bdellonemerteans are atypical in being commensals and obtaining their food by filtering the host inhalant water currents.

Food capture

Nemerteans in general feed in one of two ways, depending upon the condition of the food used. Active living prey are caught by the proboscis and killed or at least partially immobilised by its secretions before being ingested, but inert or decaying organic material is swallowed directly without prior proboscis eversion. If a suitable organism comes into range, the nemertean everts its proboscis rapidly and forcibly, and coils it tightly around the body of the intended food. The proboscis is usually aimed fairly accurately, but on occasion goes wide of its target and misses altogether. If this happens it is retracted quickly and then re-everted, this behaviour being repeated until the food has either been captured or has moved out of range. The proboscis seems to serve two purposes in this context; it mechanically prevents the prey from escaping by coiling around its body, the grip being aided by viscous secretions, papillae and rhabdites, and in many species it also produces toxic substances that kill or paralyse the prey before it is ingested. In the armed hoplonemerteans the proboscis is everted far enough to bring its central stylet to a terminal position (Fig. 8B), this then being used to inflict a wound in the body of the catch. Secretions apparently produced by the posterior proboscis epithelium (Gibson, 1970) are then pumped into the wound by muscular contractions of the powerful stylet bulb. These secretions are quite variable in their effect, depending upon the species of both nemertean and prey concerned. Some hoplonemertean toxins are clearly extremely strong, a single discharge being sufficient to kill the prey fairly rapidly (*Amphiporus lactifloreus* kills *Gammarus* within forty seconds). Other toxins seem to have only a localised effect, and a large animal must be wounded in several successive places before it becomes totally quiescent (Roe, 1970). In some instances an intended prey is only slightly or not at all affected by the proboscis secretions, due perhaps to different physiological processes which effectively confer upon them a natural immunity. It is of interest that animals coming under this category are not normally utilised as a food source by the nemertean species concerned, even when the usual supply of nutrients is depleted (Roe, 1967).

A comparable toxin effect is suggested for the anoplan species which lack a proboscis stylet (Jennings and Gibson, 1969). The simple or complex rhabdite-like barbs may be injected through the epidermis of the prey when the proboscis is squeezed tightly around it. They may serve only to assist in the proboscis grip, but it is also possible that they are responsible for the introduction of toxic substances into the prey. These could originate either from other proboscidial secretions or from the dissolution of the rhabdites themselves after their penetration. Living prey caught by species of *Cephalothrix* become inert and apparently lifeless within thirty seconds of capture, although *Clitellio* and other oligochaetes trapped by lineids are not similarly affected, and are swallowed alive.

The ingestion of food after it has been caught by the proboscis, or when already inert and not requiring the use of this organ, is then more or less uniform for the majority of species. Food is brought to the mouth by the retracting proboscis, or is approached by the worm, and apparently tested for suitability immediately before ingestion commences. During the testing period the head arches over the food and moves slowly from side to side; it is possibly during this stage of the feeding behaviour that the frontal organs are protruded and used as chemotactic sensors. If acceptable, ingestion is commenced by the distension of the mouth and lips. Vermiform prey are swallowed head or tail first, or in a U- or J-shape, small animals are taken in directly, and decaying substances or the soft body parts of large prey are sucked in as a thick fluid. The process of ingestion is frequently accompanied by peristaltic waves of the body-wall musculature passing down the body length. Swallowing thus seems to be achieved mainly by muscular activity, although ciliary action in the foregut possibly plays some part, its importance presumably depending upon the size and nature of the material being ingested. The process is assisted by the copious production of mucus or other substances, poured out from the basophilic gland cells of the buccal cavity and foregut. Where entire animals are taken as food, the criterion for ingestion appears to be diameter rather than length. Several authors have stated that animals considerably longer than the nemerteans can be regularly accepted providing their diameter is less than that of the predator (Roe, 1967, 1970; Jennings and Gibson, 1969). In anoplan forms, where the mouth is ventral, the head is arched dorsally out of the way during ingestion, but in the Enopla there is generally no need for such behaviour. The mouth in most forms is capable of extreme dilation, being pushed out around the food in the form of cupped lips.

The food is normally swallowed quite rapidly, but the ingestive stage can be prolonged for an hour or more. After being taken into the gut, it is either held in the anterior region for a short while, or is passed directly to the intestine for digestion. If a nemertean feeds on an animal longer than itself, when the gut is completely full the uningested portion is nipped off by a combination of oral contraction and enzymic histolysis. This process usually occupies several minutes.

Some nemerteans are reported to be cannibalistic or to feed on other species under laboratory conditions, but it is not known how often, or if, this happens under natural circumstances.

Unusual feeding mechanisms have so far been reported for two species. *Amphiporus lactifloreus*, after trapping its gammarid prey, does not use its proboscis to pull the food towards the head, as in other species, but releases its hold and moves forward until its anterior tip comes into contact with the crustacean. It then makes exploratory movements over and around the gammarid, apparently in search of some weak spot in the exoskeleton, such as occurs beneath the pleural and coxal plates. The head is then inserted into the prey and the stomach everted through the rhynchodaeum. If the head fails to gain entry the anterior portion only of the proboscis is everted and closely applied to the *Gammarus*. It is held in this position for one to two minutes, with the end formed into a cup- or sucker-like shape, and then withdrawn. The nemertean again attempts to insert its head, usually successfully, but will if necessary repeat the process a number of times in order to penetrate the gammarid integument. It is not known precisely how the proboscis achieves its penetration, but the use of some histolytic secretions, originating from the anterior proboscis epithelium, seems a probable explanation.

The stomach is protruded well forward through the rhynchodaeum and, with the nemertean's head inside the gammarid, is applied to the various organs and tissues of the body. These become rapidly disorganised and partially broken down, this indicating some enzymic component to the stomach secretions, giving rise to a semi-fluid mixture that is easily sucked up into the intestine. The stomach secretions are acidic in nature. *Amphiporus* will feed on dead gammarids, providing they are only freshly killed, using an identical mechanism except for the absence of initial proboscis eversion. If necessary, secondary eversion to penetrate the exoskeleton may still occur.

A feeding mechanism that is unique in nemerteans, as well as being extremely rare amongst metazoan invertebrates, is that found in the bdellonemertean *Malacobdella grossa* (Gibson and Jennings, 1969).

This is a ciliary filtering mechanism but it does not at any stage involve the use of mucus. *Malacobdella* feeds predominantly upon small algae, bacteria, protozoans, diatoms and dinoflagellates, filtering these from the water within the host mantle cavity. The foregut structure is unlike that of any other nemertean genus (Fig. 6D), particularly in the development of the very numerous and characteristic papillae. Ingestion commences with the pharynx and oesophagus compressed. With the latter kept closed, expansion of the pharynx and opening of the mouth results in water and suspended material being drawn into the body. A slight contraction of the pharynx then causes the papillae, which are suitably arranged, to interlock loosely and form a series of ciliated meshes extending the full pharyngeal length. Continued compression of the pharynx walls then forces the water back through the mouth to the exterior, but the interlocked papilliary cilia retain suspended particles. The cilia beat downwards so that small particles are transported to the papillae bases, where longitudinal tracts extend along the walls of the pharynx. Here posteriorly directed currents transport food material back to the oesophagus. The tracts merge posteriorly as the pharynx narrows towards its junction with the oesophagus and, with the latter remaining closed, particles collect into a food bolus. When a suitably sized bolus has accumulated the oesophagus is opened and the food passed into the intestine. Particles that are too large to pass down between the papillae are carried on their gently moving tips directly back to the oesophagus. The efficiency of the mechanism is such that a compact food mass can be found in the intestine only four minutes after food is administered to the host-inhalant siphon.

This filtering mechanism is impracticable for trapping the larger dietary constituents, such as *Balanus* nauplii and cyprid larvae, and *Malacobdella* catches these by means of its proboscis in the usual nemertean fashion. Proboscis retraction then brings food into the pharynx, where it is released to be carried back to the oesophagus on the papillae tips in the manner described for larger filtered particles.

Digestion and digestive enzymes

The digestive sequence in nemerteans commences with an extra-cellular acidic phase, during which the food is broken down to a size suitable for phagocytosis by the intestinal columnar cells. Digestion is then completed intracellularly in two stages, of which the first is acidic, the second alkaline. This procedure is the same for all nemerteans, but physiological differences do occur that can be related to either the diet or taxonomic position of the species concerned.

The most uniform patterns of digestion are found in anoplan

nemerteans, all those species so far investigated possessing similar processes that differ only in the duration of the individual phases. The typical anoplan sequence is represented by the heteronemertean *Lineus ruber* (Jennings, 1962; Jennings and Gibson, 1969).

Lineids ingest their food alive, relying upon the secretions of the foregut to kill their prey. During ingestion the acidophilic glands of the buccal cavity and foregut, which are rich in carbonic anhydrase, actively discharge their contents and the intralumenar pH drops to about 5·0. The basophils, which produce acid mucopolysaccharides, discharge their contents at the same time and considerable amounts of mucus can be found around ingested food. In other anoplan species, in which the prey is ingested inert or dead, the foregut acidic secretions are responsible, as they are also in lineids, for initiating the denaturation of the food and establishing the correct gut pH for extracellular proteolysis. Living prey are usually held in the foregut for a few minutes after ingestion, but inert material is passed directly back to the intestine.

The endopeptidase-producing acidophilic glands of the gastrodermis discharge as the food enters the intestine, and the gut contents become progressively more and more homogeneous. No other enzymes have been demonstrated in relation to the extracellular phase, so that in anoplan nemerteans this stage of digestion seems to be entirely proteolytic and effective at a pH of 5·5 to 6·0.

The duration of extracellular proteolysis depends directly upon the size of the meal. Phagocytosis of food particles commences about thirty minutes after feeding, and continues until the gastrodermal columnar cells are loaded with food vacuoles. Material showing endopeptidase activity may still persist in the gut lumen at this time if the amount of food ingested is large, but as intracellular digestion of the earlier food vacuoles is completed, so new ones form distally within the gastrodermal cells, this process continuing until the extracellular phase is finished. This generally occurs within twelve hours of feeding, but well before this time the lumen contents become quite homogeneous and individual animal tissues cannot be recognised, apart from such indigestible structures as polychaete setae and arthropodan exoskeletons.

The contents of food vacuoles continue to show endopeptidase activity for up to several hours after their formation, but there is no evidence to suggest that endopeptidases are secreted into them intracellularly. It is concluded, therefore, that vacuolar endopeptidase activity results from enzymic material phagocytosed along with food material from the gut lumen. The vacuoles do, however, develop a strong reaction for acid phosphatase, a less intense but still definite reaction occurring in the surrounding cell cytoplasm. The precise role of these enzymes remains undetermined, but they seem to be concerned with the maintenance of the necessary acidic pH conditions in the vacuoles for the continuance of endopeptidase activity. They may also be involved with the absorption of some products of the early intracellular digestion.

A few hours after phagocytosis is commenced exopeptidases appear in the food vacuoles, replacing demonstrable endopeptidases, and rapidly increase in amount until after some seven or eight hours the gastrodermis is loaded with vacuoles in the final stage of digestion. As intracellular digestion proceeds, the number and size of food vacuoles becomes progressively smaller. At the termination of digestion, usually within twenty-four hours, exopeptidase activity can no longer be visualised within the gastrodermal cells, this indicating that it is normally present in an inactive condition. This situation is the opposite to that found with endopeptidases, which can at all times be demonstrated within the intestinal gland cells. Similarly, strong alkaline phosphatase activity can be demonstrated within the gastrodermis at all times, although it is usually confined in its distribution to the distal regions of the ciliated cells, forming a zone of activity 2 to 4 μ wide. During the formation of food vacuoles, however, this band of activity deepens and intensifies, and at the peak of exopeptidase digestion virtually fills the cell cytoplasm and vacuolar contents. On the completion of intracellular digestion, alkaline phosphatase activity declines to the narrow distal band.

Two major functions have been suggested for the alkaline phosphatase by Jennings and Gibson (1969), one being concerned with normal cellular activities in the distal regions of the cells, such as the maintenance and renewal of cilia, the other being involved with initiating and continuing the alkaline vacuolar conditions necessary for exopeptidase activity, and with the secretion of these enzymes and the subsequent absorption of digestive products from the food vacuoles.

Lipases have so far only been directly demonstrated in specimens of *Lineus ruber* maintained on high fat-content diets (Jennings, 1962). It is probable that these enzymes are present under normal conditions, but in such small amounts that they cannot usually be demonstrated by means of histochemical techniques.

Mucus, demonstrable during the early stages of digestion, is found in the distal regions of the columnar cells in all the anoplan species so far investigated. This is derived at least in part from the foregut basophils, and is phagocytosed along with food particles. In newly formed, distal vacuoles mucus is present in abundance, but in later and older vacuoles it disappears. It is inferred from this that some intracellular carbohydrases are present in the normal digestive enzyme complement.

Greater variation in digestive physiology is found amongst enoplan species, and distinct differences occur between these and the anoplan forms, particularly with respect to the nature of the foregut secretions and the site of production of demonstrable endopeptidases.

At least two distinct types of foregut physiology exist in hoplonemerteans. *Amphiporus*, *Paranemertes* and *Tetrastemma* possess several distinct cell types, none of which exhibit carbonic anhydrase activity, although still secreting acidic substances, whereas *Prostoma* shows a stomach physiologically similar to anoplan forms in the possession of this enzyme within its acidophilic gland cells (Jennings and Gibson, 1969; Gibson, 1970). Whatever the acid-secreting mechanism involved, food taken into the gut in all those hoplonemerteans so far investigated reaches the intestine in an acidic medium, ingestion being assisted by secretions discharged from the foregut basophilic glands.

The gastrodermal gland cells discharge their contents as food enters the intestine, but the nature of their secretions has not yet been determined. The columnar cells also discharge their spherical inclusions, which they accumulate between meals, and it is these in hoplonemerteans that show a strong reaction for endopeptidases both within the cells and immediately after entering the gut lumen. As more and more spheres are discharged into the gut, where they rapidly lose their shape and become homogeneous, so the amount of demonstrable extracellular endopeptidase activity increases. There appear to be two distinct possibilities concerning the nature and function of the gland cell secretions. It is most likely that they represent proteolytic enzymes of a type not yet demonstrated, that will act in conjunction with the columnar cell endopeptidases, but it may be that their contents serve to activate the extracellular enzymes in some way as yet not determined. It seems improbable that they are involved in the production of non-proteolytic enzymes, although this possibility cannot at present be excluded.

From this point the hoplonemertean digestive processes closely follow those already described for anoplan species. Food particles are phagocytosed and for a time retain the endopeptidase activity initiated within the gut lumen, concurrent with acid phosphatases distributed around and in the food vacuoles. Although the extracellularly acting endopeptidases are synthesised and secreted from the columnar cells, there is no evidence to suggest that they are utilised at the intracellular site without first being secreted into the gut lumen.

Following this stage, endopeptidases and acid phosphatases become reduced in their activity, and are replaced by exopeptidases and alkaline phosphatases. These enzymes persist in the columnar cells until digestion is completed, when they disappear and cannot be further demonstrated until the equivalent digestive phase of a later meal. The persistent distal band of alkaline phosphatase activity, found in anoplan species, has not been reported from hoplonemerteans.

The digestive physiology of the commensal bdellonemerteans shows even further modification from the anoplan pattern, although all the changes can be correlated with the alteration to a largely herbivorous diet.

As noted earlier, food reaches the intestine in a compacted state very soon after being filtered from the sea water. The rapidity with which it passes through the pharynx and oesophagus suggests that these regions are concerned primarily with ingestion, secretions from their basophilic glands being involved in the binding together of separate particles into a discrete bolus. These secretions may also be concerned in the subsequent digestive processes, possibly as activators of enzymes released from the gastrodermal gland cells. The structure of the food bolus soon after it reaches the intestine supports these suggestions, since it consists of an inner basophilic zone intimately mixed with food particles, and an outer acidophilic region secreted from the gastrodermal glands. These two layers rapidly intermingle and disappear, the food then beginning to show evidence of digestion.

Ingested boluses are frequently isolated into definite 'pockets', clearly visible in recently fed living specimens. The 'pockets' are formed by intestinal compression anteriorly and posteriorly to the food mass, and presumably serve to permit a localised increase in extracellular enzyme concentration. Food boluses in different stages of digestion can be found within a single *Malacobdella*, and from this it is conjectured that the species feeds more or less continuously, a feature common to the majority of small-particle feeders. Food is passed down the intestine by a combination of ciliary action and alternate contraction and relaxation of the body musculature.

No enzymes have been demonstrated histochemically in the intestinal gland cells of *Malacobdella*, but there is sufficient evidence from the fate of ingested starch granules to be certain that the glands produce and secrete carbohydrases similar in their action to α-amylases. The lumen contents have a pH of 6·0 to 6·5, extracellular digestion thus being achieved in a slightly acidic medium.

The inorganic dietary components, such as shale particles and diatom cases, may in this species have an indirect and incidental function in extracellular digestion by causing abrasion and the mechanical fragmentation of the softer food components during the movement of food within the gut.

Digestion in the gut lumen is followed by phagocytosis of the products by the columnar cells, although suitably sized particles are often phagocytosed immediately after they enter the intestine. Intracellular carbohydrase activity has been identified for the species in the same way as that used for the extracellularly acting enzymes. The size limit for phagocytosis is 2 to 2·5 μ, as judged from the dimensions of intracellular inorganic material.

Intracellular digestion takes place in food vacuoles, non-specific esterases being associated with these and their surrounding cytoplasm. Esterase activity is at all times weak and mostly confined to the distal half of the cells. These enzymes are strongly inhibited by lead and silver salts, and have a pH optimum of 5·5 to 6·5; this indicates that early intracellular digestion takes place at a slightly acidic level.

Both acid and alkaline phosphatases occur in the gastrodermis, the former less frequently than the latter and being restricted to distal food vacuoles and their immediate surroundings. Acid phosphatase activity is probably similar in function to that described for other groups, and the alkaline phosphatases, in their distribution and relation to stage in digestion, are more like those of the anoplans than the hoplonemerteans.

No other enzymes have been demonstrated in the gut of *Malacobdella*, the absence of chitinases and cellulases being inferred from the continual occurrence of prey exoskeletons showing no trace of damage other than that caused by mechanical fragmentation. Particularly noticeable for their absence from this species are the endo- and exopeptidases characteristic of the free-living carnivorous nemerteans, but this variation in digestive physiology agrees very closely with the major change in diet. The digestive emphasis in *Malacobdella* is thus placed upon carbohydrases, with the demonstrable esterases probably being concerned in the intracellular degradation of proteinaceous and lipid dietary components, but the absence of demonstrable lipases may simply be due to the nature of this enzyme group in nemerteans.

The occurrence of phagocytosis in ciliated cells is of relatively rare occurrence in the animal kingdom, and its significance in the nemerteans remains uncertain. The cilia, in fact, play no part in the

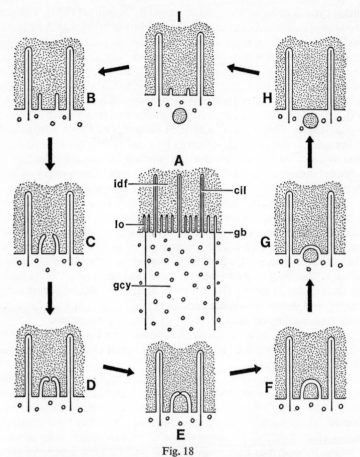

Fig. 18

Diagrammatic representation of the mechanism whereby food material is phagocytosed and passed into the intestinal columnar cells. A, Component parts of the distal region of a typical columnar cell; B–E, Outward extension of plasma lamellae; F, Fusion of adjacent lamellar tips; G, H, Food vacuole being passed back into cell; I, Phagocytosis completed, new plasmal lamellae forming. cil, cilium; gb, gastrodermal boundary; gcy, gastrodermal columnar cell; idf, intralumenar partly digested food material; lo, plasmal lamellar outgrowth.

phagocytic uptake of food, as reported by Jennings (1969). The gastrodermal columnar cells bear both cilia and microvilli at their lumen
surface, but both of these structures retain their identity and appear
in a fed specimen just as they do in the resting gastrodermis. However, the plasma membrane between these structures shows bud-like
protuberances which rapidly extend outwards into the digesting
lumen contents and develop into pseudopodia-like lamellae (Fig.
18). These lamellae vary enormously in size and may eventually reach
6 μ in length. They generally arise at distances of 1 to 5 μ from each
other and, as they extend outwards into the gut lumen, tend to curve
downwards, back towards the cell surface. Eventually the tips of
adjacent lamellae meet and fuse, trapping a mass of semi-digested food
between themselves and the cell surface. In this way a vacuole up to
5 or 6 μ in diameter, containing semi-digested food and bounded by
a single membrane, is formed and this passes back into the cell for
subsequent intracellular digestive processes. As the newly formed
vacuole moves away from the cell surface more lamellae develop and
the entire process is repeated. Thus within thirty minutes of feeding
most of the ciliated gastrodermal cells show two or more vacuoles
in their distal regions. The contents of the vacuoles become compacted and denser in appearance as they pass deeper into the cells,
presumably as water is absorbed and digestion continues.

Food vacuoles never form by invagination and vesiculation of the
plasma membrane, and the cilia and microvilli clearly play no part
in the mechanism.

The uptake of nutrient materials across the epidermis was suggested
by Fisher and Cramer (1967), who demonstrated that *Lineus ruber*
could apparently obtain amino acids and glucose in this way. Epidermal absorption of food materials has not been conclusively
demonstrated, but several nemerteans do possess epidermally situated
proteolytic enzymes that could be involved in such a process (Jennings and Gibson, 1969). This method of obtaining food, if present
in the phylum, is clearly only of secondary importance, and as such
merely plays a supportive role to the use of the gut.

Food reserves

The food reserves of nemerteans have been relatively little studied,
but in all those species so far investigated the principal food reserve
is fat, stored mainly as small globules within the gastrodermal
columnar cells, but also appearing in a number of other sites. Fat
reserves are supplemented by smaller amounts of glycogen with a
similar disposition (Reisinger, 1926; Jennings, 1960; Gibson and
Jennings, 1969; Jennings and Gibson, 1969; Gibson, 1970).

Reid (1950) investigated the glycogen metabolism of the hetero-nemertean *Micrura leidyi*. He found that during aerobic starvation there was no decrease in body glycogen content, from which he concluded that polysaccharide metabolism was not active at such a time. Conversely, when maintained under anaerobic conditions a significant depletion of the glycogen content was recorded. Under natural conditions starvation may well be encountered, but it is far more likely to be under aerobic than anaerobic circumstances. This theoretical consideration is practically confirmed by the predominance of fatty food reserves in members of the phylum, with only small and comparatively minor importance being placed upon the storage of glycogen.

Some nemerteans are notoriously capable of withstanding prolonged starvation. *Micrura leidyi* loses up to 33 per cent of its body weight in just over one month, and most species can be kept for several weeks or even months without food. Some species of *Lineus*, *Procephalothrix* and *Prostoma* can survive for more than a year, but during this time they decrease in size as the tissues are gradually absorbed by phagocytes. If the gonads are not too far advanced in development, they degenerate first, followed by many cells of the gut. The proboscis may degenerate or completely disappear, and other organs become simplified (Coe, 1943). A *Cerebratulus lacteus* 20 cm long was only about one-third of this length after four months' starvation, and *Prostoma rubrum*, kept without food except for microscopic water organisms, was only 1 per cent of its original volume after one year. During periods of starvation *Prostoma* develops a dark epidermal pigment and the epithelial ciliated cells become vacuolate (Kadis, 1951). Under the duress of starvation *Lineus socialis* normally reproduces asexually by fragmentation, but the regenerating fragments can live for nearly one year without food.

3

GENERAL PHYSIOLOGY

Our knowledge of the various physiological processes of nemerteans is, on the whole, minimal. Accordingly, certain sections of this chapter are extremely short and it will become obvious to the reader where there is abundant scope for research within the phylum.

Locomotion

Most nemerteans move quite freely over hard surfaces, burrow into sand, mud or gravel, or insert themselves into rocky crevices or beneath partially embedded rocks or boulders. Except in the larval state, relatively few nemerteans swim. Amongst the littoral and sub-littoral benthic species swimming is mostly confined to those forms in which the body is dorsoventrally compressed (*Cerebratulus, Drepanophorus, Gorgonorhynchus*), and may be performed quite actively, in contrast to the bathypelagic hoplonemerteans which either float passively or perform slow and sluggish swimming movements (Coe, 1927a, 1935; Wheeler, 1940).

Movement involves the use of both the epidermal cilia and the body-wall musculature. Ciliary locomotion, found principally in larval and juvenile nemerteans, is also used by adult worms of small size (Child, 1901; Coe, 1902b). Cilia are most commonly used in the movement of the pelagic pilidium larvae, which have ciliary bands extending around their bodies. The cilia of the marginal lappets (Fig. 17) beat in a dexioplectic manner (the effective ciliary beat is to the right of the direction of movement of the metachronal waves), setting up ripples of locomotory activity that travel around the lappet edges (Knight-Jones, 1954). The most recent study of larval and juvenile locomotion is that of Cantell (1969), who found that the pilidium swims with its cilia tuft pointing anteriorly, rotating in a clockwise direction upon its own axis. Rotational movement is temporarily suspended when the larva turns, but is resumed soon after the new course is taken up. Both the rate of rotation and speed of locomotion vary somewhat between species. Larvae of *Hubrechtella dubia*, which have a large body with comparatively small cilia, move much more slowly than do those of *Lineus bilineatus*.

The apical cilia tuft sometimes whips to and fro, at other times curves into an arc whilst distally describing an elliptical movement. Immediately prior to a course alteration, the tuft points in the new direction. Cantell interprets this as indicating that the tuft has a sensory function, although Wilson (1900) regarded it more as a 'rudder' for steering in *Cerebratulus lacteus*.

Newly hatched juveniles use their cilia in one of two ways. They either crawl along the substrate on their cilia, or else swim in a rigid position, propelling themselves by ciliary waves and not involving any muscular activity. In most cases axial rotation does not accompany swimming, although occurring in *Hubrechtella dubia*.

Juveniles frequently collide with obstacles whilst swimming. When they do so, the cilia rapidly reverse and the obstruction is avoided. In young *Lineus albocinctus* the frontal organs, which are provided with very long cilia, are projected anteriorly during swimming. When a collision occurs with an object, the organs are retracted and may thus possibly serve as thigmotactic sensors. The caudal cirrus of a creeping, possibly cerebratulid, larva is extended so far as to equal the remaining length of the body, but the reason for this is not known.

Ciliary gliding seen in young nemerteans is by no means confined to them, and some adults move in a similar manner. Pantin (1950) describes ciliary locomotion in the terrestrial hoplonemertean *Geonemertes dendyi* and relates how this type of movement is accomplished on land. The animals secrete copious amounts of mucus, particularly from the head, that is carried posteriorly by the activity of the cilia. The nemerteans thus employ their cilia to effectively 'swim' forwards inside a mucous tube, which is then left behind.

Other nemerteans creep or glide in a similar manner, but employ a series of peristaltic muscular waves passing posteriorly to achieve locomotion (Coe, 1943). The use of ciliary paralysants does not inhibit this type of locomotion, which is very similar to that described for aquatic planarians.

Muscular activity is used for all other types of locomotion, particularly the passage of major peristaltic waves, extending right around the body and not merely being confined to the ventral surface, giving an appearance similar to that seen in earthworms (Eggers, 1924). These contractions can pass posteriorly or anteriorly; even when

ciliary activity provides the motive power, the direction of movement
is governed by muscular action (Pantin, 1950), the head moving from
side to side to test the environment into which the animal is moving.
During locomotion the body shows great variation in its diameter
and shape. Many of the gregarious species become intimately
tangled with other specimens (*Cephalothrix, Emplectonema*) or, if
they are extremely long, with themselves (Fig. 1A). Many gregarious
nemerteans appear restless if isolated.

Eggers (1935) observed that the commensal *Malacobdella* moves
by a looping motion, similar to that of leeches, alternatively using the
oral aperture and the posterior sucker to adhere to the substrate
whilst the body is contracted or extended accordingly. The epidermal
cilia of *Malacobdella* beat in specific directions, inwards and forwards
dorsally, outwards and backwards ventrally, but this appears to have
no bearing upon locomotion.

In nemerteans that actively swim, the body moves easily and
rapidly in an eel-like manner, the undulatory motion being dorsoventrally orientated, at right angles to the plane of compression of
the body. When swimming the animals frequently turn on their sides
(Verrill, 1892; Wheeler, 1940). Lissmann's classical work on *Cerebratulus marginatus* (Gray, 1940) demonstrated that an animal suspended with its posterior end immersed showed two distinctive
grades of locomotory movement depending upon the external
environment. The anterior exposed half moved with typical peristaltic
muscular waves, whereas the submerged posterior portion performed
undulatory swimming actions. The inference from this work is that
normal crawling movement requires at least some degree of peripheral environmental stimulation, i.e., a thigmotactic response.

The bathypelagic forms are able to float freely, with little expendi-
ture of muscular effort, since their extensive gelatinous parenchyma
gives the body a low specific gravity, as in *Pelagonemertes* (Coe,
1927a). These hoplonemerteans are generally very sluggish in their
habits (Coe, 1935).

The use of the proboscis for locomotion has been reported for a
few nemerteans, particularly amongst the terrestrial and freshwater
genera *Geonemertes* and *Potamonemertes* (Hickman, 1963; Pantin,
1950; Moore and Gibson, 1973). This usually occurs when very rapid
movement is required (i.e., an escape reaction). The proboscis is
ejected forcibly, adhering to the substrate by its distal tip. At eversion,
which occupies less than half a second, the simultaneous contraction
of the entire body circular musculature results in an abrupt elongation of the worm to nearly twice its original length. Once the proboscis tip is firmly attached, a wave of longitudinal contraction and

circular relaxation moves forwards over the animal and its proboscis, so that it leaps forwards very rapidly, effectively retracting itself over its proboscis.

Wilson (1900) reported that *Cerebratulus lacteus* used its proboscis in a similar manner for burrowing rapidly into sand and mud.

Various attempts have been made to determine how nemertean movement is coordinated, principally by experimenting with decapitated specimens. Coe (1943) found that fragments of *Lineus socialis* containing the cerebral ganglia would continue crawling for several hours, and might move at any time during the subsequent regenerative stages. Conversely, brainless fragments remained quiescent unless mechanically stimulated. Coe concluded that normal locomotion was controlled by the lateral nerve cords in response to external stimuli, but that the initiation of movement without external stimulation depended upon the cerebral ganglia. Comparable results were obtained by Corrêa (1953a,b) who recognised three grades of locomotion based upon work carried out on decapitated nemerteans. In *Amphiporus lactifloreus*, *Lineus ruber* and *Oerstedia dorsalis* spontaneous gliding is initiated by the cerebral ganglia, the threshold of stimulation of the post-cerebral nervous elements being very much higher than that of the brain. *Emplectonema gracile* is similar to these, but has a lower threshold level for the post-cerebral body regions. In other species (*Lineus lacteus*, *Ototyphlonemertes erneba*, *O. evelinai* and *Prostomatella*) spontaneous locomotion is independent of the brain, and both the cerebral and post-cerebral thresholds for stimulation are equally low. Coe (1943) reported that pre-cerebral fragments, containing no major nervous elements, swam erratically by means of their cilia.

Many nemerteans remain immobile for considerable lengths of time, often for several days. Hickman (1963) suggests that the terrestrial *Geonemertes australiensis* does so whilst apparently waiting for prey; during this time it is enclosed in a thick covering of mucus and the absence of movement may also be related to the conservation of energy and water.

Regulation of body shape

Intimately concerned with locomotion are the mechanisms involved in the alteration of body size and shape, investigated by Cowey (1952) and Clark and Cowey (1958). The contractile and extensile powers of nemerteans have been mentioned previously; any alteration in body length is ultimately limited by the presence of the argyrophil fibrous lattice system which functions in a manner analagous to the action of lazy tongs. The mechanism can best be explained by considering a single fibre. For any given length, the helically wound fibre subtends a particular angle to the longitudinal axis of the body (Figs. 19A–C), this angle (θ) decreasing with contraction and increasing with extension. At theoretical values of 0° and 90° the volume enclosed by this system would be zero, and a simple relationship exists between body volume, length and fibre angle such that when the body is circular in cross-section (Fig. 19A), it may be either at maximal or minimal lengths (Fig. 19D). Intermediate between the two limits the volume enclosed by the system is theoretically at a maximum, but in practice remains more or less constant at all times. Thus the limits to extension and contraction of the worms are set by the points at which the actual volume is equal to the maximum that the system can permit at that particular length and value of θ, represented on the graph by the points Y and Z. Extension or contraction beyond these points (i.e., outside the curve) would require a reduction in the body volume, which is unacceptable. Maintenance of a constant body volume between the two points, where theoretically the volume increases, is achieved by an alteration in the cross-sectional area from circular to elliptical (Fig. 19B).

A theoretical extensibility can be calculated on the basis that the geodesic fibre system alone sets a limit to any alteration in length. In practice species with a low theoretical extensibility (*Geonemertes*) never depart far from the relaxed position and at all lengths are nearly circular in cross-section. In these forms it is the fibres alone that regulate changes in length. *Amphiporus*, *Lineus gesserensis* and *L. longissimus* all have an actual extensibility only a little short of the theoretical limits. *Amphiporus* and *Lineus gesserensis* attain a circular cross-section at their maximum lengths, but not at their minimum. It is concluded from this that in these species the fibre system regulates the maximum extension, but that contraction is limited by the

---

Fig. 19

Regulation of body shape. Unit lengths of a cylindrical 'worm' with circular (**A**) and elliptical (**B**) cross sections, bounded by single turns of the geodesic fibre system; **C**, A unit length slit longitudinally and flattened out; **D**, Graph to show relationship between volume contained within fibre system and angle subtended by fibres. Thin lines indicate actual volumes of various nemerteans, with superimposed thick lines showing the range over which changes in length take place; **E**, Graph to show theoretical relationship between surface area of a unit body length and the angle subtended by the fibres. **a, b**, maximal and minimal radii of ellipse; **l**, length of body unit; **r**, radius of circle; **Y, Z**, theoretical limiting positions respectively of elongation and contraction of *Amphiporus lactifloreus*; θ, angle between fibre and longitudinal axis of body. (All redrawn from Clark and Cowey, 1958)

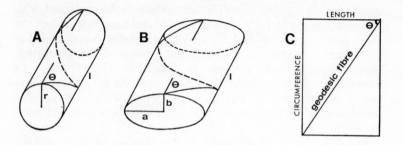

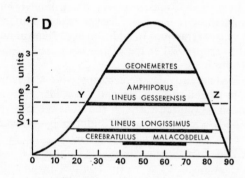

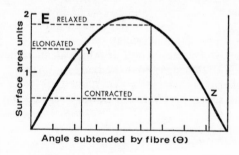

Fig. 19

minimal surface area that will accommodate all the epidermal cells, the same being true for *L. longissimus*. In these examples, when the animals are very contracted, the epidermis becomes deeply folded in an attempt to maintain surface area, but is unable to tolerate further shortening of the body towards the theoretical value (Fig. 19E). In *Lineus longissimus* the epidermal compression limits both maximum and minimum lengths, but at maximum extension the circular muscles are fully contracted and the pressures act transversely around the worm.

In other species (*Cerebratulus lacteus, Malacobdella grossa*) the actual extensibility is much less than the theoretical (Fig. 19D) but various other factors, apart from the geodesic lattice system, play an important part. These include the presence of inelastic reticulin fibres in the longitudinal muscle layers which limit extension, and similar fibres in the circular musculature which limit the circumference and hence contraction. In addition dorsoventral muscles tend to preserve the compressed shape of the body and markedly resist alterations in length.

The whole system provides nemerteans with a mechanism that combines great flexibility with strength, considerable variation in body shape being permitted through the dynamic interrelationship between the body-wall musculature and the hydrostatic pressure exerted by the body contents (Clark, 1964).

Cowey (1952) defines three recognisable lengths of nemerteans as:

(a) Relaxed, attained by a worm completely anaesthetised by magnesium ions (8 per cent $MgCl_2$ in distilled water is ideal for marine specimens).

(b) Minimum, the shortest length achieved in life by the full contraction of the longitudinal musculature.

(c) Maximum, the greatest length to which an anaesthetised worm can be passively stretched without the body rupturing.

The 'normal extension' lengths often quoted for living nemerteans are likely to fall somewhere between Cowey's relaxed and maximum definitions.

Growth and enlargement

Increase in body size in nemerteans is usually achieved by the symmetrical enlargement of tissues and structures already present, cells similar to those already present being added to the appropriate region (Coe, 1943). During this process many of the older cells become absorbed by the wandering parenchymal phagocytes, their constituents thus being returned to the general body pool of cellular material.

Elongation of the body, principally restricted to the posterior end, is accomplished by the addition of appropriate similar parts. The mechanism essentially involves a symmetrical budding in which new parts become functionally integrated with those previously existing. As the posterior regions grow so the vascular, muscular and nervous components extend into them.

In many nemerteans an increase in body size is accompanied by the multiplication of certain structures, particularly the eyes, gonads and accessory stylet pouches (Hickman, 1963). Adult lineids and certain geonemerteans, for example, have numerous eyes, but only two or four are visible in the newly hatched juveniles. As the nemerteans grow multiplication of the eyes occurs. There is no absolute correlation between head size and eye number within any one species. Hickman (1963) records that juveniles of *Geonemertes australiensis*, only about 2 mm long, have four eyes, but that this number has increased to more than twelve by the time the animals have attained a length of 5 mm.

Most nemerteans, as the body elongates posteriorly, have a rapid bilateral development of new gonads in the appropriate body region.

Both types of growth frequently involve at least some cellular replacement. The majority of the growth processes investigated have been in conjunction with regeneration, all nemerteans, with the possible exception of the bathypelagic polystyliferous hoplonemerteans, being capable of replacing lost or damaged parts to a greater or lesser degree. Regenerative capacities are particularly well known in members of the heteronemertean genus *Lineus*, and are dealt with in detail in Chapter 4.

A third type of growth is found in the armed Hoplonemertea, where the arrangement of the stylet apparatus, with the exception of the accessory pouches, precludes either of the other two growth mechanisms from being operative. In these forms size of the proboscis is more or less commensurate with that of the body as a whole, and the stylet armature is periodically replaced in its entirety. The cells which secrete the stylet apparatus become enlarged and form new parts of an appropriate size, the old stylet and basis then being discarded and replaced by the new, larger, structures. Those species in which the adults are of small size may not require stylet replacement other than that needed after damage to the existing structure, as incurred during feeding.

If parts of the body are lost through injury or amputation, similar replacements of the armature occur during the regeneration of the proboscis. In extreme cases the entire proboscis may regenerate, the discarded damaged portion then undergoing cytolysis in the fluid filling the rhynchocoel.

Growth in the genus *Carcinonemertes* is believed to be slower in the winter months than in the summer (Coe, 1902b), and a comparable seasonal growth activity may well occur in other species.

Encystment

Many nemertean species are able to form cysts or mucoid tubes under adverse conditions, although many, such as *Tubulanus*, habitually live in such tubes within sand and are capable of replenishing them in the event of loss or damage. Examples from several genera, including *Carcinonemertes*, *Lineus* and *Tetrastemma*, secrete mucous tubes only at the time of breeding, particularly during ovulation, the sheath then being discarded by the worms to form a protective case for the ova and larvae (Humes, 1942; Coe, 1943). Hickman (1963) records that the terrestrial *Geonemertes australiensis* remains stationary whilst depositing its eggs inside a mucous sheath that totally encloses the worm apart from an anterior aperture. Eggs discharged from the ovaries are lodged in the space between the epidermis and the sheath, being left behind when the nemertean crawls out of its mucous coat. The natural elasticity of the mucus seals the open anterior end, and an encapsulated egg mass remains.

Besides these types of mucoid sheaths, several species secrete enclosed cysts within which they remain coiled and, usually, inactive, often for prolonged periods of time. The freshwater *Prostoma rubrum*, under adverse or irritable conditions, secretes such copious amounts of thick mucus that it becomes totally enclosed. Contraction and hardening of the mucus result in a cyst-like structure, inside which the nemertean remains from a few days to several weeks, either lying dormant or moving slowly. If the worms are damaged prior to encystment, or unsuitable external conditions maintained for long periods of time, they may remain in their cysts until they die (Child, 1901). In this species encystment may be a device for surviving drying up of the habitat, a similar behaviour being found in the terrestrial *Geonemertes dendyi* (Pantin, 1947).

Perhaps the most studied type of cyst formation in nemerteans is that occurring during the regeneration of body fragments in such species as *Lineus sanguineus*, *L. socialis* and *L. vegetus*. In these forms small body fragments requiring extended periods of regeneration become enclosed in a firm protective cyst, remaining tightly coiled for several weeks or even months, even after regeneration is complete. The addition of further mucus from inside results in a gradual thickening of the cyst walls, waste metabolic products becoming deposited in the mucus as black granular masses (Coe, 1943). Within the cyst the nemerteans are capable of withstanding severe environmental conditions, particularly variations in salinity. When favourable circumstances return, the wall of the cyst is ruptured and the worm emerges.

Respiration

Most small nemerteans probably respire through their general body surface, as does *Prostoma rubrum* (Child, 1901), gaseous exchange occurring via the mucus covering the epidermis. In larger, slender forms respiration may be achieved in the same way, but species with bulky bodies may require special respiratory activities, as reported for *Cerebratulus lacteus* by Wilson (1900). This sand-burrowing nemertean apparently relies on rhythmic oesophageal ventilation, coupled with a complex vascular system in this part of the body, to meet at least a proportion of its oxygen requirements. Water is swallowed slowly, but expelled rapidly, a complete ventilatory cycle occupying about ten seconds. Similar oesophageal flushing occurs in *Baseodiscus* (Coe, 1906), but in this genus, with nephridial ducts discharging into the oesophagus, water movement is probably more important in the elimination of excretory materials.

Mendes (1949) states that in nemerteans the ciliation, frequent small size, elongate form and absence of a cuticle are all factors that facilitate epidermal respiratory exchanges.

Marine nemerteans kept in deteriorating conditions often move up to or penetrate the water surface; it is likely that this behaviour is at least in part concerned with respiratory needs.

Very little is known about the respiratory pigments of nemerteans, but haemoglobin has been recorded from some (Lankester, 1872; Hubrecht, 1880b; Willmer, 1970), occurring in certain blood cells (*Amphiporus, Drepanophorus, Tetrastemma*) or in the cerebral ganglia (*Amphiporus, Cerebratulus, Lineus*). Poluhowich (1970) suggests that the haem-containing pigment of the freshwater *Prostoma rubrum* may play some part in the respiratory processes. In this species the mean rate of oxygen consumption at 20°C is $0\cdot7$ μl O_2/mg dry wt/hr.

Blood physiology

The blood of nemerteans is usually a colourless fluid containing a variety of cell types (Fig. 20), including nucleated corpuscles which may contain haemoglobin, and several types of granular or agranular 'white cells' (Ohuye, 1942). The general circulation of the blood is controlled by the vessel musculature and, in some species (particularly the hoplonemerteans), valve-like flaps projecting into the vascular lumen, but much of the fluid movement is stimulated through contractions of the body-wall musculature. Flow is not normally directional, there being no heart nor equivalent region, the blood streaming forwards or backwards within the vessels and often reversing its direction regularly (Coe, 1943).

The role of the vascular fluid in nemerteans is clearly a complex one. Blood vessels from various species show an intimate relationship with such body structures as the nephridia, cerebral organs, rhynchocoel and gut, and the involvement of the blood in the various physiological processes associated with these organs appears most probable.

Amongst the enzymes reported in association with the blood system, exopeptidases of the leucine aminopeptidase type are consistently found in the endothelial and gelatinous layers of the vessel walls, less regularly in the blood plasma (Gibson and Jennings, 1967). The activity of the enzymes is independent of the nutritive state of the animals, and it is suggested that they are concerned in the circulation about the body of peptides, of use in general growth and maintenance. Calcium, a common constituent of sea water, is an activator of the enzymes, and calcium deposits in one species (*Malacobdella grossa*) may, via a phosphorylating mechanism, be concerned in the activity of the vascular proteases (Gibson and Jennings, 1969).

Histochemical techniques for the visualisation of exopeptidases can be used to demonstrate the complete blood system in whole animals (Plate 2).

Other enzymes recorded from the blood system include non-specific esterases in the endothelial layers (Jennings and Gibson, 1969; Gibson, 1970) and peroxidase in certain granular blood cells (Ohuye, 1942), but their roles have not positively been established. Peroxidases may be involved in respiration in conjunction with haemoglobin.

The blood serves to transport waste metabolites to the nephridial system, with which it often has an intimate connection, and is therefore involved in the excretory mechanisms. In this context a possible osmoregulatory role may also be suggested, although whether or not salt concentration can be regulated by the nephridial system is not known. In littoral forms capable of penetrating estuarine conditions (*Lineus, Tetrastemma*), the volume of the blood changes with alteration in osmotic concentration (Willmer, 1970), and brackish-water species possess unusually large blood spaces that Iwata (1968, 1970) interprets as indicative of an osmoregulatory function.

The vascular plugs of freshwater and terrestrial nemerteans, and

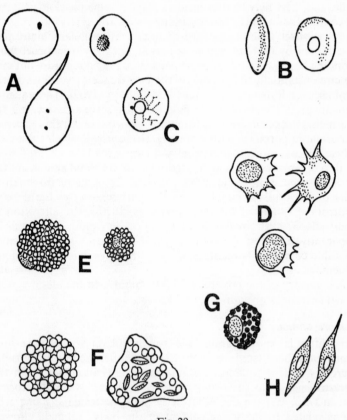

Fig. 20

Nemertean blood cell types. **A, B,** Haemoglobin-containing corpuscles of *Lineus fuscoviridis*; **C,** Haemoglobin-containing cell with basophilic reticulum; **D,** Lymphocytes; **E,** Eosinophilic leucocytes with small granules; **F,** Eosinophilic leucocytes with large granules; **G,** Basophilic leucocyte; **H,** Spindle cells. (All redrawn from Willmer, 1970, after Bürger, 1895 and Ohuye, 1942)

the rhynchocoelic villi of others, leads to interesting hypotheses concerning the relationship of the blood system with the rhynchocoel contents. The fluid filling the rhynchocoel contains various motile cells, and these may be presumed as playing some physiological role in conjunction with the proboscis apparatus. It is interesting that the proboscis itself is never vascularised; whether its cellular maintenance is achieved by nutrient materials passed, via the rhynchocoel fluid, from the blood system is at present no more than conjecture but may be circumstantially supported by the occurrence of enzymes in the proboscis endothelium (Jennings and Gibson, 1969). The aqueous constituent of the rhynchocoel fluid may also serve as a reservoir for the maintenance of osmotic balance within the body, this being of particular importance in the terrestrial *Geonemertes* which possess an unusually cavernous rhynchocoel and large proboscis.

Two other roles that may be attributed to the blood system are the possible circulation of hormones, derived from the neurosecretory cells of the cerebral ganglia, and the participation in the general hydrostatic system of the body (Willmer, 1970). Any alteration in body shape must involve local variation in volume, and this is apparently achieved through the fluid-filled and therefore compressible components of the body, the gut, rhynchocoel and blood system.

A summary of the possible roles attributable to the blood system of nemerteans is given in Figure 21.

Osmoregulation

Very little is known about the osmoregulatory mechanisms of nemerteans, but there is evidence to suggest that the protonephridial excretory system possesses a significant role (Pantin, 1947). In the terrestrial genus *Geonemertes* the nephridial system is well developed, and in *G. dendyi* the degree of activity of the terminal flame cells seems related to the water content of the body. Animals allowed partly to dry out show no movement of the flame-cell cilia, but do so soon after being immersed in fresh water. Pantin suggests that the upper part of the excretory system secretes water into the nephridial tubules, a conclusion supported by the presence of alkaline phosphatase (Danielli and Pantin, 1950), which in other groups is believed to be at least partly responsible for the mobilisation of water.

The size of the nephridial system led Iwata (1968, 1970) to the same conclusion concerning the osmoregulation of Japanese brackish-water species, and it is certainly interesting that the freshwater genera *Potamonemertes* and *Prostoma* also possess a far more extensive system of excretory ducts than is found in most marine forms

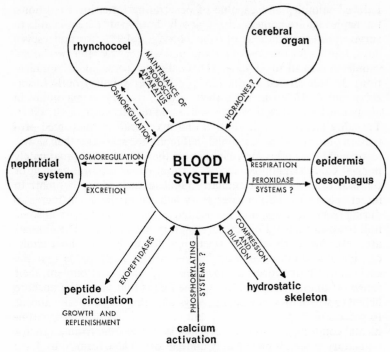

Fig. 21

Diagram summarising the importance of the blood system in various physiological mechanisms of the body. Solid arrows indicate relationships for which there is at least some experimental evidence, pecked arrows show postulated links at best supported only by circumstantial evidence.

(Moore and Gibson, 1973). In fresh water osmotic balances would result in water passing into the nemertean body, and the elimination of excess water, in these species, must be an important physiological consideration. The terrestrial hoplonemerteans, which normally live in damp conditions, are obviously able to regulate their nephridial activity according to the water content of their habitat. In these forms maximal osmoregulation will follow flooding of the environment, whereas periods of drought can be countered by the cessation of flame-cell activity when water conservation is at a premium.

Morphological evidence for the participation of the excretory system in osmoregulation is found by comparing marine nephridial development with that of terrestrial and freshwater forms. Even in littoral species like *Lineus ruber*, which is both extremely tolerant of reduced salinities and capable of penetrating estuarine conditions, the nephridia are comparatively poorly developed. *Lineus* is able to survive salinity variations of from 24·5‰ to 4·9‰ for up to seven days, 1·25‰ for twenty-four hours, and will recover after two hours' submersion in distilled water (D. G. Eason, personal communication). In less diluted media (25‰ to 30‰ salinity) both *Lineus ruber* and *L. viridis* increase their weight, reaching a maximum in about one hour when they are swollen and turgid (Lechenault, 1965). Turgidity increases with dilution and may result in the forced protrusion of the proboscis. Animals left in diluted sea water then slowly lose weight until after forty-eight hours they are again approximately normal size. Decapitated lineids, without either cerebral ganglia or cerebral organs, cannot similarly regulate their water content. In intact specimens the neurosecretory activity of the brain increases during osmotic stress, and Lechenault concludes that there is some link between this and the osmoregulatory mechanisms. The animals are nevertheless capable of tolerating, for at least a short while, considerable dilution of their body fluids. It is probable that the exclusively marine forms are isosmotic with their environment, their range of salinity toleration being one factor that limits the upshore height to which they can extend. Abyssal bathypelagic species appear to possess no nephridial system, but in the almost constant environmental conditions in which they dwell such a system, from an osmoregulatory aspect, is presumably unnecessary. These forms must have evolved different excretory pathways from the shallow-water nemerteans.

Willmer (1970) suggests that tissues other than those of the excretory system may be involved in osmoregulation. *Lineus* shows characteristic cytological responses in several groups of cells to both increased and reduced salinity, particularly those of the epidermis and foregut epithelium. Pantin (1969) comments that, at least in *Geonemertes*, there is an almost invariable close relationship between the blood system and any structure potentially involved in water secretion, and that water can be carried from the parenchyma to the organ tissues via the vascular fluid. If this is the case, then the reverse must also be true, emphasising the importance of the blood system in osmoregulation.

Excretion

The elimination of water from the body via the nephridial or other systems presumably involves the loss of certain soluble salts and metabolic products. There is some evidence to suggest that nitrogenous waste is eliminated through the nephridial ducts (Coe, 1943), but the nature of the excretory material is unknown. In the aquatic forms a reliance may be placed upon ammonia or, to a lesser extent, urea, both of which are of common occurrence in other invertebrate groups as excretory products.

Waste materials are presumably drawn from the surrounding tissues by the flame cells, and passed in a fluid medium to the longitudinal canals to be discharged through the nephridiopores (Coe, 1930a). The alkaline phosphatase present in the proximal ciliated convoluted canals and branching terminal ducts of *Geonemertes dendyi* seems likely to be concerned in the modification of the fluid passing down the nephridia (Danielli and Pantin, 1950), an osmoregulatory role discussed in the preceding section. Whether the enzyme, which occurs in marine and freshwater forms at the same site (Jennings and Gibson, 1969), is involved in salt secretion or absorption cannot at present be stated. The only other enzymes recorded from the nephridial ducts are non-specific esterases and exopeptidases, occurring in *Lineus ruber* and *L. sanguineus* respectively.

Nervous physiology

Jenkins and Carlson (1903) recorded the rate of nervous transmission in the longitudinal nerve cords of *Cerebratulus* as 5·44 to 9·0 cm/sec, but no measurements are available for any other species.

Reutter (1969a) demonstrated that in *Lineus sanguineus* the central nervous system contained primary catecholamines which might act as nerve-impulse transmitters. In the ganglionic region of the brain only a few cells are catecholaminergic, these serving mainly the peripheral nets of nerve fibres situated in the frontal organs, proboscis and foregut regions. The intraneuronal distribution of biogenic amines is, in this species, almost identical with that found in vertebrate neurones.

The role of the catecholaminergic nervous system is possibly sensory in the frontal organs, motoric in the proboscis and foregut, although some sensory participation may occur in the other two regions as well. The system appears to be important in initiating the foregut activities after regeneration (Reutter, 1969b).

Coe (1933) suggests that in *Neuronemertes* the dorsal metameric ganglia, besides serving the adjacent musculature, may also control reflex impulses in response to stimuli originating from sensory cells situated within the epidermis.

Muscle physiology

In *Lineus* some circular muscles give characteristic responses to treat-

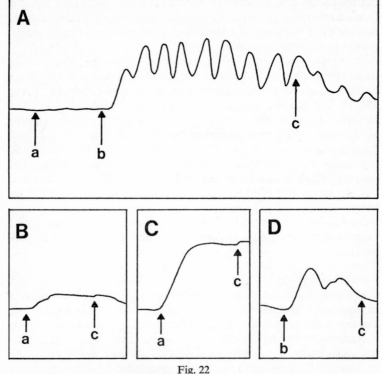

Fig. 22

Traces showing, **A**, Rhythmic type of contraction initiated by adrenaline treatment of *Lineus* longitudinal muscle, with no response to acetylcholine; **B–D**, Effects of treating proboscis longitudinal muscle with 10 γ/ml (**B**) and 50 γ/ml (**C**) acetylcholine, and 10 γ/ml adrenaline (**D**). **a**, acetylcholine administered; **b**, adrenaline administered; **c**, replacement in sea water. (All redrawn from Willmer, 1970)

ment with acetylcholine and are inhibited by atropine (Willmer, 1970), but in contrast the body-wall longitudinal musculature fails to respond to acetylcholine but reacts sharply to adrenaline by rhythmic contractions and relaxations (Fig. 22A). Other longitudinal muscles respond by a rapid contraction when exposed to oxytocin.

Acetylcholine induces arhythmic contraction in the proboscis longitudinal musculature, the degree of contraction increasing with concentration over the range 10 γ/ml to 50 γ/ml (Figs. 22B,C). Adrenalin has a similar effect (Fig. 22D) but no reaction is obtained with oxytocin. Conversely the proboscis retractor muscle contracts vigorously to oxytocin treatment.

Willmer is careful to emphasise that these substances are not necessarily those occurring in nature. The important point is that different muscles react to different stimulants, an indication that under normal circumstances each muscular system will possess its own excitatory combination.

The main body-wall musculature comprises two different fibre orientations that are essentially antagonistic, and this suggests that either a nervous control predominates or that hormonal substances or other chemical transmitters are employed. Thus the two muscle layers may respond to different stimulants, or differently to the same stimulant.

Cholinesterase, which may be involved in the physiological processes of the musculature, is known to occur in some nemerteans (Smith, Jackson and Prosser, 1940; Bacq, 1947; Kamemoto, 1957) in both specific and non-specific varieties. Ling (1969a,b) has demonstrated the occurrence of the enzymes in several nerve tracts, particularly those leading to the proboscis. Kamemoto (1957) records that *Prostoma rubrum* has an extremely high cholinesterase activity (454 to 899 units) compared with other invertebrates, and more than one hundred times greater than that found in *Cerebratulus lacteus*.

Other enzymes recorded in various muscle tissues include alkaline phosphatase and esterase (the latter possibly a manifestation of cholinesterase activity), but the roles of these enzymes are not yet known (Jennings and Gibson, 1969; Gibson, 1970). The nature of the naturally occurring muscle stimulants has still to be discovered.

Enzymes of other tissues

Various enzymes are known to occur in tissues other than those already described. The epidermis of many species variously contains esterases, acid phosphatases and exopeptidases, mostly restricted in their distribution to a narrow distal zone (Jennings and Gibson, 1969; Gibson, 1970). Nemertean epidermal secretions are often strongly acidic (McIntosh, 1868, 1873–4), but in a few forms are alkaline. The secretions, which are mucoid and viscous, presumably serve at least two other functions besides facilitating locomotion. These are the maintenance of a moist body surface, and the probable protection against potential predators (Bacq, 1937). Crabs, extremely voracious littoral carnivores, will only feed on nemerteans if very hungry, and characteristically clean the mucus off with their claws prior to ingesting them (D. G. Eason, personal communication). This suggests that the mucus is at least irritating or distasteful, although in some species, such as *Paranemertes peregrina*, toxins similar to those found in the proboscis occur throughout the epidermal regions (Kem, 1971). Epidermal enzymes may, therefore, be involved in the synthesis and secretion of these substances.

Possibly a more plausible explanation for the role of these enzymes is that they are concerned with the absorption of simple nutrient materials from the environment. Fisher and Cramer (1967) showed that *Lineus ruber* could take up amino acids and glucose across the epidermis, possibly by means of a pinocytotic mechanism involving the epidermal microvilli, although comparable structures in the gastrodermis are known not to be involved in the uptake of food material (Jennings, 1969). Under normal conditions any absorption of nutrients across the epidermis is likely to be of secondary importance to the functions of the gut, although it is possible that certain substances, contained in solution in the surrounding water, are selectively absorbed via the epidermal enzyme systems. It is not known whether these enzymes are present in the epidermis of terrestrial species.

Poluhowich (1968) found that epidermal granules in *Prostoma rubrum* gave a slight positive reaction for peroxidase activity. The density of the granules increases during starvation; this is interpreted as possibly indicating that during starvation food reserves or bodily tissues are being utilised, unused or degraded portions being deposited within the epidermis. Since the animals also have haemoglobin the granules may represent the haem group of the molecules, stored or excreted whilst the protein component is incorporated into the general physiological processes. The involvement of the peroxidases in epidermal respiration cannot be excluded.

The roles of enzymes in other parts of the body in certain species cannot at present be explained, although suggestions can be made. In one species (*Prostoma rubrum*) carbonic anhydrase, an enzyme associated with the production of acidic secretions, occurs in glands of the proboscis, and in the common British species *Amphiporus lactifloreus* some or most of the anterior proboscis secretions are involved in an extracorporeal histolytic process that forms a part of this species' feeding mechanism (Jennings and Gibson, 1969).

Esterases occur in the cephalic furrows of *Lineus ruber* and the cerebral glands of *Paranemertes*, and exopeptidases in the cephalic slits of *Lineus sanguineus* (Jennings and Gibson, 1969; Gibson, 1970). The foregut epithelium of several species also contains demonstrable esterases, not apparently involved in any digestive capacity. Since the foregut epithelium and epidermis are embryologically homologous, comparable functions for their enzymes may be suggested.

Tissue homogenates of *Lineus ruber* contain enzymes capable of transforming acid Δ-aminolevulin into porphobilinogen (Vernet, 1968), but the precise site of activity is not known. These enzymes may play some part in the biosynthesis of haemoglobin for use in the vascular and nervous systems.

Neurosecretion

The occurrence of neurosecretory cells in the cerebral ganglia and lateral nerve cords was discussed by Lechenault (1962) for lineids, although the role of their products is far from clearly established. Suggested functions include osmoregulation (Lechenault, 1965) and hormone production (Scharrer, 1941; Willmer, 1970), although Bierne (1964, 1966) suggests that in female *Lineus ruber* there does seem to be some relationship with the onset of sexual maturity.

Lechenault (1963) reported that the neurosecretory material was a glycoprotein rich in basic and sulphur-containing amino acids, whilst Bianchi (1969b) showed that the neurosecretory cells could be characterised by the presence of cystein, tryptophan and other 3-indolyl derivatives. None of the cells appear to contain lipids demonstrable in paraffin sections, and Bianchi found that in only a few could he detect a carbohydrate component.

An evolutionary parallel between the cerebral organs and the vertebrate pituitary complex has been suggested by Scharrer (1941), who points out that the four developmental stages recognisable in nemerteans, the most primitive occurring in the Tubulanidae, the most advanced in the Lineidae, show a correlation between their anatomy and level of involvement with the blood system. Scharrer suggests that these stages may represent an evolution from an exocrine to an endocrine gland, and Willmer (1970) points out that the dorsal epithelium of the buccal cavity in *Lineus ruber* shows staining affinities similar to those in the anterior lobe of the rat pituitary hypophysis. Bianchi (1969a) found that the storage area of neurosecretory material in the neuropil may function as a primitive neurohaemal structure, analogous in function with the neurohaemal organs of other animals. The principal sites for neurosecretory cells are the mediodorsal regions of the brain, around the cephalic clefts.

There is thus some evidence to suggest that components of the cerebral organs, cerebral ganglia and foregut are equivalent to an endocrine precursor of more advanced animal forms, but the establishment of an endocrine function is still far from definite. Balfour and Willmer (1967) found that the entire pharyngeal region, especially the anterior portion, accumulated radioactive iodine from the surrounding medium, although other tissues, such as the epidermal mucous glands and proboscis, were less regularly involved. Iodine taken up may be stored for some time, but only a small proportion occurs bound in thyronine compounds. A relationship exists between the amount of iodine accumulated and thyronine formed with the salinity of the medium, more being picked up from concentrated sea water than from a diluted solution, even when the amount of available iodine is the same in both cases (Fig. 23). This suggests that the foregut epithelium accumulates iodine in relation to the general salt balance of the environment, and that iodine and chloride ions are not necessarily alternatives to be taken up according to their availability. Considerable variation in the degree of accumulation occurs between different specimens, but this is not yet understood.

Toxins

The toxin systems of nemerteans have been investigated for only a few species. Bacq (1936) isolated 'amphiporine' and 'nemertine' from *Amphiporus* and *Lineus* respectively, and investigated their pharma-

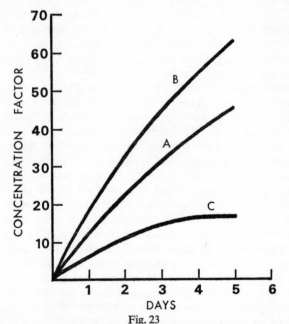

Fig. 23

Graph to show uptake of radioactive iodine by foregut epithelium of *Lineus ruber*. **A**, Worms kept in artificial sea water; **B**, Worms kept in artificial sea water to which is added 1 g NaCl/100 ml; **C**, Worms kept in artificial sea water diluted to 50 per cent salinity with distilled water. (Redrawn from Balfour and Willmer, 1967)

cological effects. 'Amphiporine' at a 1:500,000 dilution rapidly stimulates contraction of the frog *rectus abdominis*, and is exactly matched by a 1:400,000 dilution of nicotine, a related alkaloid (King, 1939). Bacq found that the chemical, which is apparently contained in the muscular tissues and not localised in the stylet apparatus, taken from *Drepanophorus crassus* excited and then paralysed cat ganglion cells isolated and artificially perfused, but that the effect was only temporary. A large dose of 'amphiporine' is lethal to crabs.

'Nemertine' has a similar effect on crabs, and represents some 0·3 per cent of the worm's body weight. The degree of toxin activity varies between species and is absent from many related forms.

Bacq concluded that the two substances were not true venoms, but served more as protective chemicals rather than as actively used poisons.

A more recent study has shown that the toxin of the hoplonemertean *Paranemertes peregrina* is the tetrahydropyridine anabaseine (Kem, Abbott and Coates, 1971). This substance resembles 'amphiporine' in possessing nicotinoid characteristics, but does not have any direct effect upon nerve fibres of crustaceans. Kem (1971), in finding that the bulk of the anabaseine (some 70 per cent of the total) occurs in the body wall, supports Bacq's (1937) interpretation of the defensive role of toxins.

Within the proboscis apparatus only minimal amounts of the toxin are found in the central and posterior regions. From a theoretical point of view it is in these parts of the organ that secretions, of use in conjunction with the stylet apparatus, should originate. In *Paranemertes*, however, approximately 90 per cent of the total proboscis anabaseine occurs in the anterior proboscis (about 7 per cent of the dry tissue weight), particularly high concentrations appearing in the mucus adhering to the papillae. This somewhat anomalous situation cannot at present be satisfactorily explained.

Kem (1971) has investigated the toxins of other species in addition to *Paranemertes peregrina*. Anabaseine also occurs in *Amphiporus angulatus*, *A. lactifloreus* and *Tetrastemma worki*, but never to the extent that it is found in *Paranemertes*. The most toxic species so far investigated are *Lineus socialis* and *L. viridis*; extracts of these forms cause convulsions similar to those initiated by anabaseine treatment, but are irreversible in their effect. *Lineus* toxin is a polypeptide neurotoxin, and corresponds to 'nemertine'. The toxin of *Amphiporus ochraceus* may be the 6-piperideine isomer of anabaseine.

Temperature effects

The effects of temperature upon nemerteans have been but barely studied. Hickman (1963) observed that *Geonemertes australiensis* could be kept for six months in a vivarium provided the temperature did not exceed 23°C, but that only a few degrees warmer (27°C) usually proved fatal to them.

The common littoral *Lineus ruber* is able to withstand a wide range of temperatures for at least seven days (0°C to 30°C). Immersion in sea water at 40°C and above proves immediately lethal, but it is not clear whether death is caused by the temperature itself or shock resulting from a sudden temperature rise. There is some evidence that *Lineus* can withstand freezing for short periods of time, an ability of potential benefit in intertidal environments.

The thermal death point for *Lineus ruber*, when the temperature is raised by 1°C in five minutes, is in the order of 36°C to 36·5°C (D. G. Eason, personal communication).

Responses to stimuli

Responses shown to various kinds of stimuli may differ greatly even in closely related species (Coe, 1943). *Lineus socialis*, for example, in response to strong tactile, electrical, chemical or thermal stimulation contracts into a tight spiral, whereas *Lineus ruber*, whilst still contracting, does so without coiling. In both species the proboscis may be everted so violently that it ruptures its retractor muscle and even tears free from its anterior insertion. This behaviour is also readily seen in irritated *Geonemertes chalicophora*.

The various sensory organs of the body have been described in Chapter 1; despite much careful experimentation our knowledge on both the roles and mechanisms of the various structures in relation to external stimuli is still minimal.

Although adult *Emplectonema* move directly towards light sources (Eggers, 1924), most nemerteans are negatively phototactic and move promptly away in search of shade (Coe, 1943; Kirsteuer, 1967b). They are, accordingly, mostly active at night or on dull overcast days. In contrast, the free-swimming larvae are positively phototactic (Colgan, 1916; Coe, 1943), this situation being commonly found in littoral invertebrates with a pelagic larval phase in their life cycles.

The detection of light is not confined to the eyes, since decapitated specimens, in general, behave in a manner very little different from whole worms (Child, 1901; Gontcharoff, 1953). Conversely, *Lineus lacteus* loses its photonegative orientation after the removal of its brain and eyes (Corrêa, 1953a). Gontcharoff suggests that the extra-ocular light response is localised in certain cells of the central nervous system and not in the epidermis, and that the eyes in fact are mostly not of major importance in controlling light orientation. Ling's

(1969b) work supports this hypothesis in demonstrating that eyeless *Lineus ruber* only react in the normal manner to light beams directed on the cephalic region; those illuminated on other parts of the body do not show comparable responses. This further suggests that any nervous involvement in extraocular photodetection is possibly confined to the cerebral ganglia and may be absent from the lateral nerve cords. Corrêa's (1953a) observations agree with this inference.

Amongst intertidal species that regularly emerge during the daytime to feed when the tide is out, such as *Paranemertes peregrina* (Roe, 1970) and *Tubulanus annulatus* (Gibson, unpublished), it is possible either that the adults normally respond positively to light, or that the usual negative reaction is modified by tidal or other conditions. *Lineus ruber*, when lit from below, responds in the usual negative manner except when confined to small water volumes. In these circumstances the phototaxis is apparently upset by thigmotactic senses (Gontcharoff, 1952). *Lineus* seems to possess a dorsoventral gradient to light responses; when lit from above its reactions are more or less directional (therefore, strictly, a photokinesis), but when lit from below it more or less wanders at random.

The eyeless genus *Ototyphlonemertes* has no demonstrable phototactic sense, and fails to respond to any variation in light intensity (Corrêa, 1948), a similar situation being found in the commensal *Malacobdella* (Eggers, 1935).

Nemerteans commonly show quite vigorous responses to chemical changes in the water, particularly to potentially harmful substances. The cephalic grooves of *Cerebratulus* and *Lineus* open and close rapidly in response to mild stimuli, the body then reacting by contracting away from the excitatory source. More intense stimulation, with stronger chemical concentrations, may cause such severe contractions that rupturing follows. The proboscis may be repeatedly everted and, in extreme cases, breaks free from the body (Coe, 1943). Ling (1969b) showed that extreme increase or decrease in salinity induces *Lineus ruber* to close its cephalic furrows and cerebral gland openings.

D. G. Eason (personal communication) found that in *Lineus* different parts of the body reacted differently to stimulation by dilute hydrochloric acid or sodium hydroxide. The cephalic region is sensitive to pH ranges of 1 to 5 and 10 to 14, but in the pH 6 to 9 range some individuals fail to respond. None of the worms reacted within the values 6·1 to 8·2. Similar responses to salt crystals were found by Ling (1969b), *Lineus* being capable of detecting salinity variations of 20 mM quite accurately.

Under laboratory conditions fouling of the water by accumulation of carbon dioxide or other deleterious substances often stimulates the animals to crawl above the water level. They will remain there until they die unless the water is changed. Child (1901) found that *Prostoma rubrum* tended to aggregate in regions of high oxygen concentration.

Eggers (1935) was unable to detect any evidence of chemotactic response in *Malacobdella*, but suggests that some form of response may be shown by senses mediated via as yet undetermined structures. Riepen (1933) found that two worms, on meeting, behaved antagonistically towards each other and related this to their normally solitary mode of life. Certainly when a number of *Malacobdella* do occur together within a single host they are mostly very small and well distributed in the bivalve mantle cavity (Gibson, 1967).

The chemical detection of food, which in some species can be achieved over distances of several centimetres, is discussed in Chapter 2.

Most of the body surface reacts to tactile stimuli, but the response is more marked in the extreme anterior and posterior regions. Decapitated worms respond in a manner almost identical with that of entire animals. Coe (1943) states that the proboscis, besides its offensive and locomotive functions, serves also as an extremely sensitive tactile organ.

In *Lineus ruber* tactile stimulation results in partial contraction away from the point of contact, all regions of the body responding equally. A continuous pressure is slightly more effective than spasmodic stimulation (D. G. Eason, personal communication). Eggers (1935) found that mechanical irritation of *Malacobdella* caused contraction or extension depending upon whether the stimulus was applied anteriorly or posteriorly respectively. Loosening of the sucker often followed stimulation, but with a delay, so that it is impossible to state with certainty that this is a behavioural response.

In the few species investigated for thigmotactic reactions, all are strongly positive (Child, 1901; Corrêa, 1948), showing a strong tendency to crawl into narrow crevices, between sand grains, or beneath stones and algae. In the absence of mineral objects normally solitary species will become intimately entangled with other specimens. Movements in response to thigmotactic stimulation appear not to be orientated. Child (1901) suggests that *Prostoma*, once encysted, may be prevented from leaving because of a very strong positive thigmotaxis.

Ototyphlonemertes shows a positive but temporary reaction to geotactic stimulation when their orientation is disturbed. When left undisturbed, specimens may crawl uphill, downhill or along a level and no geotactic activity can be determined (Corrêa, 1948). It is possible in these forms that they may migrate up and down the shore in rhythm with tidal fluctuations, following the water level.

No form of geotactic response can be found in *Malacobdella* (Eggers, 1935).

Nemerteans seem to have a strongly developed positional sense, rapidly and persistently righting themselves if they are overturned. This behaviour is equally well performed by decapitated specimens and in isolated heads. The only nemertean genus to possess statocysts is *Ototyphlonemertes* (Corrêa, 1948, 1950, 1953b), and it is presumed that the sense is mediated via the general nervous system (Coe, 1943).

Eggers (1935) postulated that the commensal *Malacobdella*, living in the mantle cavity of bivalve molluscs and therefore subjected to ciliary water currents, would show a distinctive rheotactic response to current direction. This, however, he was unable to demonstrate, but suggested that it may be present as a highly complex sense corresponding to the complicated currents found within the host.

Pigments

Exceedingly few analyses have been carried out upon nemertean pigments, and the commonest so far characterised seem to be variants of the carotenoid group (Lönnberg, 1933, 1934; Fox, 1954), particularly from the brightly coloured species. In *Nectonemertes mirabilis* the red-orange colouration is due to an acidogenic carotenoid ester, related to but not identical with astaxanthin, containing no carotene hydrocarbons and with a single absorption peak of 490 mμ. A similar pigment is reported from the terrestrial *Geonemertes hillii* (Pantin, 1969), who lists for other species a porphyrin or porphyrin-like substance (*G. novaezealandiae*) and a more insoluble, possibly melanin, pigment (*G. pantini*). Melanin is suggested as the pigment in the cerebral organ of *Drepanophorus* (Bürger, 1895).

Vernet (1966) discovered that in *Lineus ruber* three distinct but coexisting pigments appeared. One, rather like melanin, occurs in the eyes, and two, ommochrome and porphyrin, are found in the dermal tissues. The metabolism of porphyrins, investigated by Vernet and Gontcharoff (1971), is such that regenerating pigment cells are biochemically differentiated before they begin to show as morphologically distinct structures.

The significance of the brilliant colouration and patterns found in so many nemerteans is not understood, but it seems most probable that they are protective devices against potential predators. It is perhaps significant that of the two intertidal species known to feed openly during daylight hours and when the tide is out, *Paranemertes peregrina,* with a deep purplish-black colour, is strikingly obvious against its muddy background, and *Tubulanus annulatus* has a characteristic pattern of white longitudinal and transverse stripes and bands against a bright orange body. Conversely, the more drably coloured species remain hidden and emerge only at night and when submerged. Professor A. J. Cain (personal communication) suggests that an analogy may be drawn between nemertean colouration and that seen in other invertebrate groups, where characteristic and easily recognisable patterns serve as a protective mechanism through predatory learning. This inference could clearly explain an otherwise puzzling situation.

Bioluminescence

The only bioluminescent nemertean species so far reported is *Emplectonema kandai* (Kanda, 1939). The animals luminesce brilliantly in response to mechanical, thermal, electrical or chemical stimulation, the light, which is whitish green, appearing all over the body surface but lasting for only one or two seconds.

Luciferin and luciferase have not been demonstrated for the species, but the light source, contained in epidermal cells but absent from the tip of the head, is not apparently of bacterial origin, as it is in many bioluminescent organisms. The nature of the luminescence is so far not known.

Parasites

Parasites are of common occurrence in nemerteans and can be found in most parts of the body, particularly the intestinal lumen and general parenchyma. Unicellular parasites are most frequently encountered, although metazoan forms such as trematodes and nematodes have been found on occasion.

In the gut lumen gregarines are often present in large numbers (Kölliker, 1848; Roe, 1971; Moore and Gibson, 1973), but infection by ciliates is also known. *Malacobdella grossa* has been reported with two ciliate gut parasites, *Thigmophrya annella* (Fenchel, 1965) and *Orchitophrya malacobdellae* (Jennings, 1968). In most cases gut parasites appear to cause little or no damage to the gastrodermis, merely swimming freely amongst the host gut contents. Gregarines showing *in tandem* syzygy are reported from the blood system of the freshwater *Potamonemertes percivali* (Moore and Gibson, 1973), but it is not certain whether this is the same species as that occurring in the gut of the same host.

Sporozoans too have been found in the gut lumen (Punnett, 1901c), but more commonly appear in other tissues of the body.

Brinkmann (1917) found sporozoan-like structures in the eggs of *Nectonemertes primitiva* and the brain and general parenchyma of *Parabalaenanemertes fusca*.

In most cases the long-term effects of the parasites are not known, but *Haplosporidium malacobdellae*, a sporozoan infecting *Malacobdella grossa*, is known both to eventually kill the host and, if infection occurs before the onset of sexual maturity, to cause parasitic castration (Jennings and Gibson, 1968). There is no suggestion of acquired immunity with increase in age. A related form, *Haplosporidium nemertis* from *Lineus bilineatus* (Debaisieux, 1920), does not apparently have a similar effect.

An isolated report of hyperparasitism was made by Vinckier, Devauchelle and Prensier (1970), the microsporidian *Nosema vivieri* living in the tissues of monocystid gregarines from the gut of certain nemerteans.

4

ASEXUAL REPRODUCTION AND REGENERATION

Some degree of regenerative ability is probably a characteristic of all nemerteans, with the possible exception of the bathypelagic types, as a means of replacing lost or damaged body parts. Many species possess bodies that, being slender and fragile, are very susceptible to mechanical damage, and the arrangement of the muscular and nervous systems in most is such that comparatively mild stimuli result in contractions violent enough to cause fragmentation, particularly of the posterior extremities (Coe, 1934a). The possession of mechanisms for the repair of damaged tissues is therefore of prime importance to nemerteans, although for the majority operative only in the replacement of posterior structures.

Spontaneous fragmentation and asexual reproduction

The greatest powers of regeneration in nemerteans are found in the few heteronemertean species that possess an asexual reproductive phase, accomplished through spontaneous fragmentation, in their natural life cycles (Zhuravleva, Korotkevich and Korotkova, 1970).

Fig. 24

Asexual reproduction and regeneration. A–C, *Lineus vegetus*; A, Complete worm coiled and partially contracted; B, Same individual fragmented; C, Regeneration of fragments into miniature worms of normal proportions. Two fragments have encysted (x); D, Posterior regeneration of fragments taken from various parts of the body in *Zygeupolia rubens*. A, regeneration complete to give miniature worms, B, posterior regeneration only with anterior formation of vesicles (v), C–F, posterior regeneration only. B–F eventually die and disintegrate. cc, caudal cirrus; cg, cerebral ganglia; f, foregut; i, intestine; m, mouth; oe, oesophagus; sm, stomach. (A–C redrawn from Coe, 1931; D redrawn from Coe, 1934a)

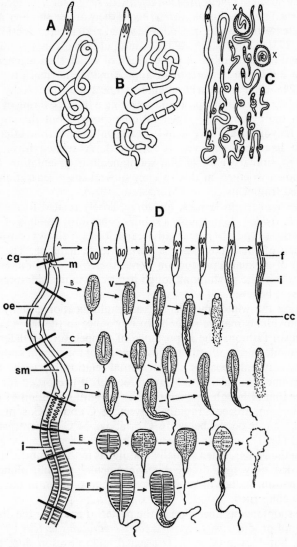

Fig. 24

In these species (*Lineus pseudo-lacteus, L. sanguineus, L. socialis* and *L. vegetus*) asexual fission, occurring in the warmer summer months, alternates with a winter period of normal sexual reproduction (Coe, 1930d, 1931; Gontcharoff, 1951). Adult but sexually immature worms fragment into a number of pieces, each of which then regenerates to form a miniature nemertean of approximately normal proportions (Figs. 24A–C). Particularly small fragments may encyst by coiling themselves into a tight spiral and secreting a thick covering of mucus. Within the cyst normal regeneration proceeds, but the completed worm may not leave its mucoid sheath until some time after regeneration has been achieved (Coe, 1931). A related form, *Lineus nigricans*, similarly encysts after spontaneous fragmentation, but it is not known whether, in this species, subsequent regeneration of the enclosed fragments occurs (Monastero, 1928).

Other nemertean species, including closely related forms (*Lineus ruber, L. viridis*), do not show a similar behaviour, and it is of interest that in these types the regenerative capacities are always considerably less than of those possessing a natural asexual phase. The phenomenon of asexual reproduction is not therefore of common occurrence within the phylum, although it may be more widespread than at present believed.

During the winter months asexual multiplication usually ceases, and individuals mature sexually to reproduce in the more common nemertean fashion. The importance of temperature in determining which mode of reproduction takes place is not for certain known and, in fact, may only possess a coincidental relationship with other, at present undetermined, conditions. Coe (1929b) recorded for *Lineus vegetus* that although regeneration, an integral condition of asexual reproduction, proceeds rapidly between 18°C and 20°C, some reduction in the rate occurs below 16°C or above 24°C. In contrast Gontcharoff (1951) found that *Lineus sanguineus* fragmented as successfully in the winter, when sexually mature, as in the summer. It can be concluded only that different environmental stimuli, alone or in combination, induce or inhibit asexual reproduction in the various species concerned.

The spontaneous liberation of the posterior end after the discharge of genital products was reported for *Cerebratulus lacteus* by Wilson (1899), but does not constitute asexual multiplication since the discarded posterior part degenerates, only the anterior end regenerating its missing portions.

The mechanisms involved during asexual reproduction in the regeneration of fragments are essentially the same as those occurring in other forms of regeneration and, according to Coe (1934a), closely agree with aspects of normal embryological development. Three distinct stages may be recognised, the closing and healing of the wound, the formation and proliferation of a regenerative bud, and the differentiation and regulation of tissues to dimensions commensurate with the size of the regenerated worm.

Wound closure

Whether an injury is caused artificially or naturally, the initial reaction is for closure of the wound to occur by the firm contraction of the body-wall musculature next to the damaged surface. In cases in which the injury involves rupturing of the rhynchocoel epithelium, the proboscis is usually so forcibly ejected that it breaks away from its attachment and is discarded. Contraction of the musculature draws the epidermal structures around the wound together, causing compression of the underlying tissues which seals off the cavities of the blood vessels, gut and rhynchocoel (Coe, 1929b). In *Lineus sanguineus*, and probably in other species that fragment spontaneously, wound closure is anticipated before breakage occurs, the epidermis, dermis, body-wall muscles and lateral nerve cords separating at the sites of fragmentation (Reutter, 1967). Partial epidermal invagination results in these sites being externally visible as paler coloured constrictions encircling the body.

Whether encystment occurs subsequently or not, such large amounts of mucus are secreted that the entire fragment as well as the damaged surface usually becomes enclosed in a viscid sheath. In most instances wound closure takes place rapidly, but pre-cerebral pieces may fail to heal and quickly degenerate.

Wound healing is accomplished by cellular migration (Nusbaum and Oxner, 1910, 1911a,b, 1912; Coe, 1929b, 1934b), and involves cells originating from both the parenchyma and the proximal regions of the epidermis. Undifferentiated epidermal cells, termed neoblasts by Coe (1929b), are primarily responsible for the development of the thin flattened epithelium that quickly forms over the wound, spreading from the damaged margins (Fig. 25A). Neoblast migration is not apparently polarised, since wound healing occurs with equal facility on all damaged surfaces irrespective of orientation, provided the original fragment contains a sufficient number of epidermal cells. In pieces much less than half as long as their diameter, surface injury generally fails to heal since there are insufficient epidermal cells to cover the wound without serious disruption of the existing epithelium.

Once a new epidermis has formed over the wound, cellular proliferation supplemented by the incorporation of additional neoblasts

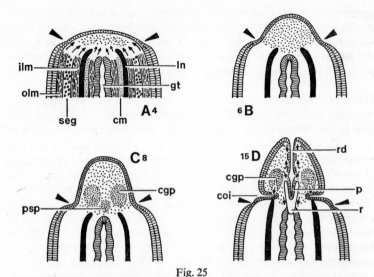

Fig. 25

Regeneration. **A**, Cellular migration in wound closure; **B**, Formation of anterior blastema (stippled) prior to organ differentiation; **C**, Localisation of prospective organs within blastema; **D**, Blastema with differentiated primordia of cerebral ganglia and regenerating proboscis apparatus. All figures refer to *Lineus socialis*, numbers beside figure letters indicating the post-operative time in days. **cgp**, cerebral ganglia primordium; **cm**, circular musculature; **coi**, cerebral organ invagination; **gt**, gut; **ilm**, inner longitudinal musculature; **ln**, lateral nerve cord; **olm**, outer longitudinal musculature; **p**, proboscis; **psp**, proboscis and proboscis sheath primordia; **r**, rhynchocoel; **rd**, rhynchodaeum; **seg**, subepidermal glands. (All redrawn from Coe, 1934b)

migrating to the area results in the epithelium assuming a more cuboidal or columnar appearance (Figs. 25B,C), the margins of the injured region becoming even more constricted (Coe, 1934a). Cilia appear on the free surface as in a normal epidermis. The migration of neoblasts and their subsequent differentiation into epidermal components is essentially similar to that found in the regular replacement of worn-out epidermal cells.

Parenchymal phagocytes, together with those arising from other body tissues, ingest damaged and degenerating cells and cellular debris, transferring any nutrients so obtained to the differentiating epithelium.

A narrow cavity develops between the wound and its newly formed epidermis, the fluid filling this containing large numbers of amoeboid phagocytes which migrate in from original tissues. Some transport nutrient materials with them that are derived from the resorption of undamaged but non-essential body parts. These cells are particularly abundant immediately beneath the new epidermis, and it is from them that the greater proportion of regenerated tissues, excluding the epidermis, are constituted. Although very closely associated with the final stages of wound healing, they are more strictly to be considered as involved with the regenerative processes.

The migratory cells produce a secretion that contains minute fibrils, and these constitute a primitive parenchyma. This ensures that the wound is firmly closed, at the same time protecting the underlying tissues from infection by bacteria or other organisms. From this condition (Fig. 25A) the fragment is prepared for the regeneration of its lost tissues and organs.

Regenerative buds and regeneration

The extent to which regeneration can take place depends both upon the species and body region concerned. Most nemerteans appear capable of posterior replacement, but very few can regenerate structures anterior to the damaged region, and then only if some of the original cerebral or nervous tissue is retained by the fragment.

Motile phagocytes of parenchymal origin appear everywhere amongst the developing tissues. They can be distinguished from other parenchyma cells by their larger size and granulated cytoplasm. Many are vacuolate, and several contain pigment granules.

In the regeneration of anterior structures an accumulation of undifferentiated migratory cells beneath the newly formed epidermis gives rise to a projecting regenerative bud or head blastema (Coe, 1929b, 1930e, 1934a,b; Gontcharoff, 1951; Reutter and Sasse, 1970), composed of what are then variously termed regenerative, formative, blastocyte or neoblast cells. It is these that eventually give rise to all the structures of the new anterior end apart from the epidermis, cerebral sense organs, foregut epithelium and mouth which are derived from development of the original epidermis. The blastema is a bilaterally symmetrical structure, its plane of symmetry apparently becoming fixed coincident with the determination of organ primordia. A true blastema does not develop in *Prostoma graecense*, the only hoplonemertean species so far reported as able to regenerate a new head, including the cerebral ganglia, from a fragment whose anterior end, including the brain, has been removed (Kipke, 1932), although it is only able to do so if the regenerating fragment contains the proboscis and some portion of the original foregut.

The blastema (Figs. 25B,C), which is clearly demarcated from the original tissues, is at first regionally undifferentiated and occupies the entire region beneath the newly formed epidermis, essentially filling in the subepithelial cavity formed during wound healing. It rapidly becomes organised and, at the time when the organ primordia are being determined, is in possession of full regenerative ability. Cellular determination seems to depend entirely upon where cells happen to be situated at the time when differentiation commences, since blastema cells not confined to primordial regions are capable of further migration, whereas those constituting the primordia are not. The precise time and mechanisms involved in the initiation of determination are at present unknown. For some time after the primordia have become established (Fig. 25c) additional cells migrate into the blastema from the original tissues, supplementing the primordial cells which are themselves multiplying by mitotic division.

In the regeneration of a completely new head the first blastema structures to differentiate are the paired primordia of the cerebral ganglia. At first they consist simply of clusters of ovoid or amoeboid cells, similar to but more closely grouped than those of the remaining blastema regions. Mitotic division within the primordial limits increases cell density, and some addition of extra blastema cells also occurs. The primordial tissues soon become converted into recognisable ganglia cells, which then send out nerve fibres that become grouped into a distinctive fibrous core, corresponding to the central fibrous region of the cerebral ganglia. The cores of the ventral ganglionic lobes then grow posteriorly (Fig. 25D) to join up with the remnant of the appropriate original lateral nerve. If the fragment only contains a single nerve cord, fusion with the ventral lobe occurs on that side, whilst proliferation and elongation of the opposite core and ganglion cells results in the posterior growth of a new lateral nerve.

Replacement of the proboscis and its associated structures can occur in one of two ways, depending essentially upon whether its loss is accompanied by damage to other tissues or not. In the regeneration of a completely new head, including the cerebral ganglia, development commences with the formation of a proboscis primordial region in the mid-line between the differentiating brain lobes (Fig. 25c). This rapidly becomes organised, growing posteriorly to form a thin-walled fluid-filled sac that soon comprises the rhynchocoel and sheath (Fig. 25D). At the anterior end of this sac further primordial tissues form an accumulation of regenerative cells, initially of similar appearance to those developing into the rhynchocoel epithelium but proliferating more rapidly to produce a posteriorly elongating proboscis bud (Coe, 1934b). This soon becomes attached to the posterior end of the developing rhynchocoel by means of fibrous processes. The constituent layers of the proboscis are later differentiated at the same time as nerves extend into the organ from the newly regenerated cerebral ganglia.

Loss of the proboscis without damage to other anterior tissues can occur either spontaneously, after exposure to severe irritation, or experimentally, the organ being forcibly everted and then carefully torn free from the body. In either case rupture of the cephalic attachment at the front of the brain leaves a short slender basal stalk whose lining is continuous with that of the rhynchodaeum. This stalk is capable of extremely rapid growth (Gontcharoff, 1951) and, assisted by proliferation of the rhynchodaeal epithelium, quickly regenerates a new proboscis.

In *Lineus ruber*, *L. sanguineus* and *Amphiporus lactifloreus* regeneration is preceded by the development of a blastema composed of epidermal and parenchymal derivatives (Bierne, 1962b). Unless the original fragment is of comparatively large size, when the developing proboscis may grow back into the existing rhynchocoel, the old proboscis sheath degenerates and is removed by parenchymal phagocytes, a new structure growing posteriorly in its place. If none of the original proboscis apparatus is present, space for the developing structures is still provided through phagocytic activity, and essentially involves the absorptive clearing of a pathway in the dorsal parenchyma as the regenerating tissues proliferate posteriorly.

Complete replacement of the head requires the development of a new rhynchodaeum. This originates in the blastema as a narrow epithelial tube that grows forwards to join up with an invagination of the anterior epidermis (Fig. 25D), so as to form a continuous tube. The dual origin of the rhynchodaeal epithelium is reflected in the mixing of epidermal and parenchymal cells. In these instances activation of the rhynchodaeal epithelial development occurs before that of the rhynchocoel and, in fact, the epidermally derived blastocytes that comprise a major part of the rhynchodaeum appear to possess some activatory effect upon the proliferation of the rhynchocoelic parenchymal constituents (Bierne, 1962b).

The cerebral sense organs, of epidermal origin, are derived from a pair of lateral tubular invaginations that develop at the boundary between the blastema and the original tissues of the fragment (Fig. 25D). Posterior outgrowth of the cerebral ganglia later provides them with their nervous ganglionic elements. The cephalic grooves similarly arise from longitudinal folding of the newly formed epidermis, extending anteriorly along the margins of the head from the cerebral canal aperture.

A new gut can be regenerated even if none of the original structure is retained by the fragment (Dawydoff, 1910; Nusbaum and Oxner, 1910, 1912), although there is some disagreement as to the source of the formative cells. Coe (1934a) regards the initial part formed as the anterior region of the new intestine. In *Lineus socialis* the first indication of the new gut is an accumulation of phagocytic and other parenchymal cells near the centre of the recently regenerated tissues. As more and more cells become aggregated, a fluid-filled space, the potential lumen, appears in their midst. This cavity soon becomes surrounded by regenerative cells which at first form an undifferentiated sheath, but later develop the typical gastrodermal structure enclosing a regularly defined lumen. The anterior end retains its blastemic formation for some time, the regenerative cells either forming irregular lobes projecting into the lumen, or moving freely within the intralumenar fluid. As the body of the regenerating nemertean begins to elongate posteriorly, the hind end of the new intestine becomes somewhat constricted, later joining up with a terminal epidermal invagination that forms the anus.

At the front end of the newly formed intestine a layer of regenerative cells, distinctly demarcated from the gastrodermal tissues, differentiates into an epithelium that represents the primordium of the foregut. This first appears as a crescentic anterior outgrowth of the intestine, but later becomes tubular and more elongate through the incorporation and proliferation of additional parenchymal cells. Soon after, the ventral epidermis at the junction between old and new tissues forms a slender invagination that comprises the new mouth. Internal enlargement of the invagination gives rise to the buccal cavity, which later extends posteriorly to join with the cavity of the foregut primordium. The foregut, therefore, is composed partly of epidermal and partly of parenchymal tissues.

Epidermal proliferation and the formation of the blastema, as well as the differentiation of the cerebral organs, foregut and proboscis, are histochemically characterised in the same way (Reutter and Sasse, 1970). At the start of regeneration the tissues have a high glycogen content, but this decreases with further development at the same time as there is an increase in both RNA synthesis and glucose-6-phosphate dehydrogenase activity. This is interpreted as indicating that in areas of active regeneration glucose formed by glycogenolytic processes is metabolised, via the pentose-phosphate shunt, to be utilised in the synthesis of nucleic acids which are of prime importance in the regenerative mechanisms.

Regeneration of the posterior tissues never involves the development of a true blastema, and the regenerative bud can usually give rise only to those tissues normally present to its posterior. The bud

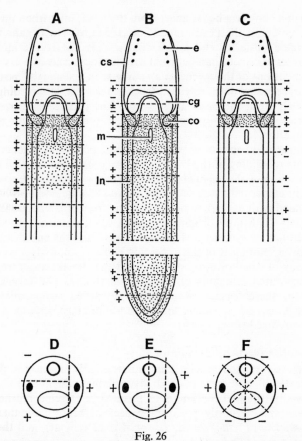

Fig. 26

Regenerative capacities of body parts. A–C, Diagrams indicating the regenerative capacities of groups of lineid species when their bodies are cut transversely; **A**, *Lineus pictifrons* and *L. rubescens*; **B**, *L. socialis* and *L. vegetus*; **C**, *L. ruber*; **D–F**, Diagrammatic transverse sections of *L. socialis* showing dependence on a part of the lateral nerve cord for complete regeneration. **cg**, cerebral ganglia; **co**, cerebral organ; **cs**, cephalic slit; **e**, eye; **ln**, lateral nerve cord; **m**, mouth; +, complete regeneration and regulation; ±, occasional regeneration; −, regeneration incomplete. Stippled regions in **A–C** represent regions of the body where both anterior and posterior regeneration occur. (**A–C** redrawn from Coe, 1934a; **D–F** redrawn from Coe, 1930e)

can form a complete body, apart from the head, only when the fragment consists of the head alone (Coe, 1934a,b). The organs of the regenerated posterior end are derived principally from undifferentiated reserve cells of parenchymal origin. These cells first lay down the foundations of the regenerating parts, then become differentiated and incorporated into specific structures as the latter proliferate posteriorly from the original fragment. The processes of posterior regeneration are basically similar to those of normal posterior growth by elongation.

In those lineid species possessing extreme powers of regeneration, complete rebuilding of the body can be accomplished by even a very small portion of the original worm, provided the fragment possesses sufficient epidermal tissue to permit wound closure and healing, and is provided with at least some part of the original lateral nerve cord (Figs. 26D–F). Minute fragments regenerate comparatively slowly, but eventually a symmetrical blastema forms and the new head regulates the development of the remaining body. Commonly a complete new body of small dimensions is formed that breaks away from the unincorporated remnants of the original tissues. The regenerative ability of these species is such that discarded tissues may then develop a second blastema, and an additional nemertean is eventually produced (Coe, 1930e).

Comparative regenerative abilities

Even closely related species may differ greatly in their ability to regenerate different parts of the body. Amongst the lineids a gradation can be found between those possessing extreme regenerative potential and naturally performing asexual reproduction (*Lineus pseudo-lacteus, L. sanguineus, L. socialis, L. vegetus*), and those in which not even two new worms can be regenerated from the original one (*L. ruber, L. viridis*). A single adult *Lineus socialis*, for example, can be repeatedly cut into so many pieces that more than one hundred new worms are regenerated, and in this species the only criterion governing regeneration appears to be the presence of a small part of the original nerve cord (Coe, 1929b, 1930e, 1934a,b). The related *Lineus pictifrons* forms an interesting intermediate between the two regenerative extremes found in the genus (Coe, 1932), in that complete regeneration is possible in any fragment cut between the cerebral ganglia and the mid-foregut region, whereas only a small proportion of pieces taken more posteriorly successfully replace all the lost tissues. In front of the lateral nerve cords only anterior regeneration occurs, and behind the middle of the foregut replacement is normally confined only to posterior structures. A summary of the range of regenerative potentials in the genus *Lineus* is illustrated in Figs. 26A–C.

Examples from other genera show varying degrees of regenerative ability, but none approaches the extreme powers shown by the asexually reproducing lineids. The heteronemerteans *Cerebratulus lacteus, Euborlasia nigrocincta* and *Micrura leidyi* all show some evidence of spontaneous fragmentation, but never in association with any form of asexual reproduction. None appear able to regenerate anteriorly, but posterior replacement is usually complete providing the original fragment contains both at least some part of the intestine and intact cerebral ganglia (Coe, 1934a). The heteronemertean *Zygeupolia rubens* shows remarkable posterior regenerative ability (Fig. 24D), and segments of practically any part of the body will rapidly regenerate the caudal cirrus. In addition, both anterior and posterior wound healing occurs rapidly, but there is never any evidence of blastema formation. Accordingly anterior regeneration does not occur, and a complete worm can only be formed by posterior proliferation of a severed head. Other parts of the body, even though replacing the caudal cirrus, and often part of the gut, eventually die and disintegrate.

In these species the cellular regenerative elements are principally derived from parenchymatous tissue, although some involvement of the gut epithelium may also occur.

The regenerative capacities of hoplonemerteans have been little investigated, and *Prostoma graecense* appears to be the only species capable of regenerating a completely new head (Kipke, 1932). Other species are able to replace anteriorly only those structures situated in front of the cerebral ganglia (Coe, 1934a). In *Tetrastemma vittatum* the rhynchodaeum regenerates first from an accumulation of parenchymal blastocytes, followed in turn by the oesophagus, proboscis and epidermis, all of which at least in part are anterior to the brain (Sandoz, 1965). The cerebral sense organs can also be replaced in part by invaginations of the new epidermis fusing with those portions of the organs remaining in the original tissues.

Posterior regeneration, essentially by normal growth, is somewhat limited, and is usually only complete if the amount of tissue removed is small.

Although there is comparatively little direct evidence to support the suggestion, it seems probable that the proboscis can be replaced by most, if not all, nemerteans. Of all the body structures this organ, by the nature of its participation in food capture or, less frequently, locomotion, is potentially susceptible to damage or loss, and it seems unlikely that an otherwise complete animal at this level of organisation would be unable to regenerate such a necessary part.

Regulation of regeneration

In all types of regeneration the migratory formative cells are initially multipotent (Coe, 1934a,b) and do not assume the characters of specific tissues until after differentiation has occurred. Prior to regenerative determination the only inherent differences apparent between both epidermal and mesodermal (parenchymatous) formative cells can be related to their original embryological development, in that epidermal derivatives will never differentiate into structures that are normally of mesodermal origin, and vice versa.

When regeneration occurs in association with a tissue already partly represented in the original fragment, determination appears to be influenced by some stimulus emitted by the existing functional tissues. The fate of regenerative cells in these instances therefore depends entirely upon the nature of the tissue with which they become associated.

The same is not true when complete organs are lost and need to be regenerated, such as in head amputation, and blastemic differentiation must be governed by other, at present undetermined, factors.

Coe (1934a) suggests that the movement of regenerative cells is polarised, so that their direction of migration is governed by their position relative to that of the cerebral ganglia. If this control was retained even in the absence of the brain, then the inability of most nemertean species to regenerate anteriorly could be explained simply by the absence of suitably polarised cells from the remaining tissues. That the regeneration of pre-cerebral structures can be effected by many species lends support to this hypothesis, particularly since transverse cuts made through the cerebral lobes are usually, if not invariably, fatal to both fragments. This is true even for those lineids with extreme regenerative abilities. Presumably the species able to regenerate complete heads including the cerebral ganglia either possess anteriorly polarised formative cells situated behind the brain or are in some way able to regulate the direction of polarisation as required.

The dependence upon the presence of at least some portion of the lateral nerve cords for complete regeneration suggests that the organisational centre is confined to these structures (Coe, 1932). The cut end of the nerve cords apparently liberates an agent which activates the undifferentiated and dormant formative cells, transforming them into an active state, and it is the bipolar migration of these that permits regeneration (Coe, 1934b). The different regenerative abilities shown by closely related species may therefore be dependent upon differences in the distribution of this agent within the nerve cords. Species possessing asexual reproductive ability, and therefore extreme regenerative potential, presumably have the effective agent distributed throughout the body length, whilst those with lesser powers of replacement have a correspondingly reduced organisational distribution.

This theoretical interpretation is both supported and extended by Tucker's (1959) studies on *Lineus vegetus*. She found that both anterior and posterior regeneration involve a series of differentiation centres extending sequentially from the site of initial regeneration (the blastema or posterior bud respectively), such that each centre is both regulated by that immediately preceding it, and influences the degree of differentiation shown by that immediately following it. Each centre thus possesses its own characteristic differentiation pattern; it is unable to attain levels of tissue organisation that have already been reached because of the inhibitory effects of the centre that has differentiated immediately before it. This interpretation is based upon evidence that extracts of head blastemas totally inhibit the anterior regeneration of any part of the body, whereas extracts of the posterior bud only inhibit posterior replacement. Similarly extracts of the mid-body regions prevent blastema formation only in those fragments taken posteriorly to the extracted region but not anteriorly, whilst the reverse is true for posterior regeneration. The nature of the inhibitory reagents has not been determined, but they occur in the supernatent moiety of centrifuged tissue homogenates.

Tucker also found that the more closely related two species are, the greater is the likelihood of regenerative extracts from one inhibiting differentiation of the appropriate body region in the other. Thus cephalic homogenates from *Cerebratulus californiensis* inhibit blastema formation in *Lineus vegetus*, but those from *Amphiporus formidabilis* do not. Tail or mid-body extracts give comparable results. Clearly even in those species incapable of anterior regeneration (*Amphiporus, Cerebratulus*), the inhibitory substances are still present. Since the blastema forms a self-determining system comparable to that of an early embryo, with its constituent cells potentially capable of differentiating into any of the requisite tissues (Coe, 1934b), it can be further concluded that the inhibitory and stimulatory substances are present in embryonic nemerteans, but that the latter, effective for anterior regeneration, are only retained in adults of the few species capable of complete anterior regeneration.

The size attained by the newly regenerated worm depends upon the amount of material present in the original fragment, for old tissues are gradually resorbed by parenchymal phagocytes to provide the nutrients for the proliferating cells of the regenerative buds. Coe (1943) regards these buds, particularly the blastema, as parasitic upon the original tissues. If the fragment is so large that not all of its cellular material is required for regenerative formation, the remaining tissues and organs eventually become remodelled into similar parts of smaller size, in harmony with the reduced dimensions of the newly formed nemertean.

Grafting

Experimental work involving the grafting of a body part from one individual on to that of another is necessarily difficult in animals of this type because of their tendency to coil up, and is frequently unsuccessful, even in those nemerteans with good regenerative capacities. Coe (1930e) records that fragments of *Lineus socialis* split longitudinally are as capable of complete regeneration as are portions of the body cut transversely, providing the criteria governing the replacement mechanisms for the species are met. Longitudinal fragments from different individuals can, on occasion, be induced to fuse more or less completely, although more often each component of the graft retains its own individuality and regenerates independently of its partner.

Whether or not tissue fusion does occur seems to be entirely a matter of chance, and is not apparently dependent upon the polarities of the two partners being coincident. Usually any fusion that does take place is only temporary. Each partner acts as an independent entity in the development of its blastema, and ultimately the two halves separate to form a pair of small worms that regenerate in the normal individual way. Very occasionally the degree of tissue fusion is such that later fragmentation of the graft gives rise to three individuals, one of which comprises tissues derived from both of the original donor specimens. This composite nemertean in all respects regenerates as a normal worm.

Bierne (1967a) investigated the effects of grafting fragments from two closely related species together, basing his work upon five varieties of *Lineus*. He found that whilst the chimera produced to some extent remained viable, each component regenerated according to its own phenotype and was not apparently influenced by the development or nature of its partner. In most cases the heterospecific regenerates were destined to failure, but there appears some possibility that future work may enable the growth of these artificial 'species' to be prolonged.

Twinning, organ duplication and monstrosities

The duplication of organs during normal embryological development is of occasional natural occurrence, but can be obtained more frequently during regeneration experiments, particularly when sagittal splitting is repeated during the replacement processes (Coe, 1930e). A single longitudinal cut, extending from the tip of the snout to just behind the cerebral region, will, if left undisturbed, heal by a combination of lateral and anterior regeneration. However, additional cutting, and posterior extension of the split, performed before tissue reorganisation is completed, will often result in pairing of the original blastema, each controlling the regeneration of its own anterior structures. In this way a two-headed nemertean can be produced which will later fragment spontaneously to form three small worms, two from each of the twinned head regions, and one from the single posterior region that separates off at the junction of the two heads.

Less extensive sagittal splitting may permit fusion of the cerebral commissures even when a second cut is made before initial healing is completed. In these instances, however, when splitting is performed in conjunction with cephalic amputation, some effect is shown on posterior regeneration and specimens with twinned tails can be formed. Even when the twin tails are of the same size and appearance, they survive only for as long as it takes the proboscis apparatus to be regenerated, for the rhynchocoel, during its posterior elongation, usually grows only into one or the other tail region and rarely or never bifurcates to extend into both. That tail portion into which the rhynchocoel is growing then seems stimulated to further growth, with the result that the arhynchous tail eventually becomes resorbed and the twinned nature is lost.

If two heads of unequal size are formed by asymmetrical longitudinal cutting, or one of the paired regenerated heads is later amputated close to the junction with its partner, the larger or remaining head will inhibit further renewal of its twin and assumes dominance for the entire body. In the first instance the smaller head becomes gradually resorbed, a process that is not hindered by amputation of the larger head providing dominance of the latter has been established. An amputated twin head can be successfully replaced if the level of amputation is some way anterior to the junction of the twinned tissues.

Natural twinning can occur through embryonic duplication following the partial fusion of two developing ova. Coe (1943) gives as an example the development of paired proboscides, each complete with a central stylet and basis and three (instead of the normal two) accessory stylet pouches, in *Oerstedia dorsalis*. Other naturally occurring examples of twinning are apparently produced after injury through duplication during regeneration. Several instances are known of *Cerebratulus* or *Lineus* with forked tails (McIntosh, 1906; Gontcharoff, 1951), and Gontcharoff (1949) found a solitary *Lineus sanguineus* with neither cephalic nor caudal regions in which regeneration was taking place laterally. Eventually the regenerating worm became detached and grew as a normal individual. In most cases examples with duplicated body parts can only be regarded as monstrosities, and presumably have correspondingly reduced chances of survival.

Eunemertes (= *Nemertes*) *echinoderma* can be quite commonly found in nature with paired proboscides (Gontcharoff, 1958a), although in these instances, since the larger and presumably older proboscis is detached and clearly in a degenerating condition whilst the smaller is attached and showing signs of development, it does not appear to be so much a case of true organ replication as of replacement in conjunction with normal growth. Experiments on regeneration in this species (Gontcharoff, 1958b) showed that when the anterior attachment of the proboscis is severed, a regenerative bud develops within a few days, and this grafts itself on to the cut end of the proboscis to give rise to a complete organ. No other cases are known of proboscis replacement or repair occurring in this way.

5

SEXUAL REPRODUCTION AND EMBRYOLOGY

Reproductive seasons

In most nemertean species sexual reproduction takes place during the warmer summer months, although several examples are known in which the breeding period is either variable, depending upon the geographic location, or is confined to the winter season. Examples of summer-breeding species include *Amphiporus ochraceus*, *Carcinonemertes carcinophila*, *Lineus torquatus*, *Micrura akkeshiensis*, *Procephalothrix simulus*, *Tetrastemma candidum* and *Tubulanus punctatus*, whilst *Amphiporus lactifloreus*, *Cerebratulus lacteus*, *Lineus ruber*, *L. socialis* and *L. viridis* reproduce sexually at other times of the year (Coe, 1899b, 1902a; Iwata, 1957a, 1958, 1960a; Gontcharoff, 1960). A summary of the breeding seasons of several European and North American species is given by Bürger (1897–1907). The only species at present known to reproduce at all times of the year is the entocommensal bdellonemertean *Malacobdella grossa*, although superimposed upon the breeding cycle there are biannual peaks of fecundity that can be correlated with the availability of the species' phytoplanktonic food (Gibson, 1968).

At the onset of the reproductive season it seems probable that species which are not normally gregarious become aggregated to ensure successful mating. Coe (1901) found nearly fifty ripe *Micrura alaskensis* entangled together in coarse muddy gravel, and several other usually solitary forms can be found in increased numbers. Exceedingly little is known about the factors governing spawning in nemerteans, but the presence of a ripe adult is capable of stimulating the activity of others without the necessity for physical contact. Wilson (1900) reported that male and female *Cerebratulus lacteus* maintained in different parts of the same aquarium discharged their genital products simultaneously, and several instances are known of gravid females apparently delaying ovulation until the approach of mature males. Bierne (1966, 1967b) demonstrated that the sexual maturation of *Lineus ruber* was governed by some cerebral neuro-crine system, although in other experiments (Bierne, 1964) he found that decapitation of females resulted in precocious maturation. The nature of the secretion involved appears to differ between the sexes, male maturation being governed by an androgenic factor that can be used experimentally to mascularise the ovaries (Bierne, 1967b, 1968, 1970a,b). A gonado-inhibitory factor produced by the cerebral ganglia exerts a strong inhibitory effect upon sexual development (Bierne, 1970a,c). From this it may be conjectured that the secretion of a neurocrine or other chemical substance from one sex or the other, liberated into the water via the cerebral sense organs, could stimulate the discharge of genital products by the partner, as wit-nessed for *Cerebratulus* by Wilson (1900). Such a secretion would therefore be pheromonal in its nature.

In general palaeo- and heteronemerteans mature and lay all their eggs either simultaneously or within a few days at most. Several, particularly amongst the genera *Cerebratulus*, *Euborlasia*, *Lineus*, *Micrura* and *Tubulanus*, are capable of producing extremely large numbers of eggs, often up to fifty per ovary. The ovulatory potential of a large specimen is well illustrated by reference to Coe's (1899b) estimate, that a *Cerebratulus lacteus* five feet in length could, in a single season, ripen and lay up to 250,000 eggs. The 22-foot-long specimen recorded by Verrill (1892) could, on this basis, yield one million or more eggs.

In contrast, most of the enoplan species mature only one to three large-sized eggs per ovary at any one time, but since the reproductive period is usually extended over several weeks (Coe, 1943) the total productivity may still be quite large. McDermott (1966), for example, found that a pair of females of an unidentified but *Carcinonemertes*-like species produced 175 egg sacs containing approximately 14,000 embryos.

When the number of eggs produced is small, they are usually of comparatively large size, those of *Tetrastemma caecum*, as an example, being two-thirds the diameter of the body (Coe, 1905a). Similarly, Humes (1942) found that in *Carcinonemertes* unfertilised eggs could reach diameters of 58 μ, whilst the gravid females were as small as 98 μ in width. The eggs of anoplan nemerteans, on the other hand, are generally small in comparison to the body size, particu-larly when produced in enormous numbers. Diameters of 100 μ to 200 μ are not uncommon, but the worms themselves may be several

millimetres wide (Coe, 1905a; Iwata, 1957a, 1958, 1960a). The largest
eggs so far reported for pelagic nemerteans are those of *Dinonemertes
investigatoris*, which may be 2·5 mm in diameter (Coe, 1926).

Gonad maturation

At the approach of the breeding period certain of the follicular cell
nuclei lining the ovarian epithelium increase in size rapidly, forming
the germinal vesicles of the future eggs (Coe, 1905a). Other cells
become filled with yolk globules, and essentially each egg is formed
through the aggregation of yolk cells around a germinal vesicle, the
whole structure then becoming enclosed by a thin to thick egg membrane (Fig. 27A). At one side, and always opposite the excentric
germinal vesicle, the egg membrane is extended as an attachment
stalk linking the egg with the ovarian lining.

In males the maturation of the testes similarly commences with
development of the follicular lining cells. These give rise sequentially
to the spermatogonia, spermatocytes, spermatids and spermatozoa,
and one or more of these stages may be seen within the ripening testes
(Fig. 16E). The filamentous sperm (Fig. 27B), termed a nematosperm

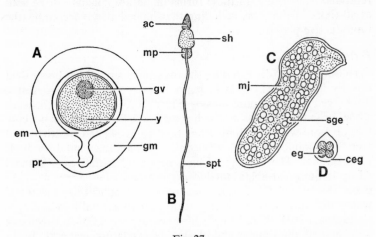

Fig. 27
Sexual reproduction. **A**, Egg of *Micrura akkeshiensis* immediately after
liberation; **B**, Mature spermatozoan of *Micrura akkeshiensis*; **C**, Gelatinous egg mass of *Lineus ruber*; **D**, Individual egg capsule of *Lineus
ruber*. **ac**, acrosome; **ceg**, capsule containing eggs; **eg**, egg; **em**, egg
membrane; **gm**, glutinous membrane; **gv**, germinal vesicle; **mj**, mucoid
jelly; **mp**, middle piece; **pr**, protuberance; **sge**, group of eggs in capsule;
sh, sperm head; **spt**, sperm tail; **y**, yolk. (**A**, **B** redrawn from Iwata,
1958; **C**, **D** redrawn from Hyman, 1951, after Schmidt, 1931b, 1934)

by Iwata (1957a, 1958), consist of a variably shaped head bearing an acrosome, a middle portion, and an extended flagellate tail. Even closely related species may possess ripe sperm of quite different form. The pattern of microtubules within the tail does not necessarily correspond to that found in cilia of the same species (Henley, 1970).

The formation of the gonopores, which does not take place until sexual maturity is attained, is achieved via the outward growth and extension of certain follicular lining cells. The ripe eggs and sperm are then squeezed from their gonads through contractions of the body-wall musculature, although, particularly amongst the bathypelagic forms, muscle fibres specifically associated with the gonads play a significant part in the process. In several instances, especially amongst those hoplonemerteans producing large eggs, ovulation may involve rupturing of the general body wall, but this does not necessarily lead ultimately to the death of the parent following reproduction. Coe (1905a) suggests that whilst many species are probably annual, others, including examples of *Amphiporus*, *Cerebratulus* and *Drepanophorus*, live for several years and pass through several successive reproductive seasons. In these forms immature gonads can be seen at all times of the year, including that immediately succeeding the reproductive period.

Fertilisation

Fertilisation is mostly external, the eggs and sperm either being shed directly into the sea or, as in many hetero- and hoplonemerteans, discharged into gelatinous masses (Figs. 27c,D) that are laid beneath stones or in the burrows in which nemerteans live. In the second case two or more worms of both sexes become associated together, and secrete a common mucoid envelope into which the genital products are liberated. The number of eggs deposited within each mucoid string is quite variable; *Amphiporus ochraceus* lays forty or more (Coe, 1899b), but *Geonemertes australiensis* produces egg masses of widely differing sizes containing from four to several hundred eggs (Hickman, 1963), there being a fairly close correlation between egg number and the dimensions of the encapsulating material. Iwata (1960a) reports that the loose jelly-like substance enclosing the eggs of *Procephalothrix simulus* usually dissolves fairly soon after spawning.

When the genital products are deposited within a mucoid capsule there is a likelihood of sperm entering the ovarian gonoducts and fertilising the eggs internally. Coe (1905a) considers this to be a primitive form of copulation. Usually the eggs are laid shortly afterwards, but several cases are known in which development proceeds to the blastula stage or, less commonly, as far as the differentiation of the pelagic larva, before ovulation occurs (McIntosh, 1873–4; Coe, 1902b; Humes, 1942).

After mating the adults leave the mucoid egg cases, which then close and partially collapse to form semi-solid masses in which the eggs are embedded.

Amongst the bathypelagic species more elaborate means of ensuring fertilisation have been evolved that can be correlated with the sparse distribution of the adults. Mature male *Nectonemertes* possess a pair of cephalic tentacles (Fig. 1c) that are used to hold the females during mating and, in these forms, the development of the spermaries (testes) into protrusible penes is indicative both of internal fertilisation and a true copulatory behaviour. Adult male *Phallonemertes* similarly possess long penes, but these project laterally from the cephalic region at all times after the nemerteans have matured sexually (Figs. 16F,G). Males of another bathypelagic species, *Plotonemertes adhaerens*, possess a pair of sucker-like adhesive organs which are presumed to serve for holding the females during reproduction. Coe (1926) observes that breeding in the abyssal species probably occurs only once in their lives, the adults dying after mating, and that under the stable environmental conditions at such depths sexual reproduction is not likely to be seasonal. There is thus all the more reason for nemerteans of these types to develop comparatively elaborate reproductive mechanisms in order to ensure successful fertilisation.

Internal fertilisation also occurs in the terrestrial hoplonemertean *Geonemertes agricola*, which is usually or strictly ovoviviparous. Development to young worms, complete except for their gonads, takes place almost entirely within the ovaries (Coe, 1904, 1939). Internal fertilisation with subsequent partial or complete ovoviviparity is also believed to occur in certain species of *Prosorhochmus* and *Poikilonemertes*. The only heteronemertean so far reported to bear live young is *Lineus viviparus*.

Amongst the hermaphroditic species (*Amphinemertes, Campbellonemertes, Coenemertes, Dichonemertes, Geonemertes agricola, Poikilonemertes, Potamonemertes, Prosadenoporus, Prosorhochmus, Prostoma, Tetrastemma*) self-fertilisation must be a possibility, and has been demonstrated under laboratory conditions at least for one form. If, and how regularly, this occurs in nature is not known. Some hermaphrodites, particularly in the freshwater genus *Prostoma*, are apparently protandric; the male and female phases may be separated temporally, or an overlap may occur during which true hermaphroditism is present. So far as is known, no nemertean species exhibit protogyny.

Development

Many investigations have been made upon the various aspects of nemertean embryology. These include studies on the palaeonemerteans *Tubulanus* (Dawydoff, 1928; Iwata, 1960a), *Cephalothrix* (Smith, 1935), and *Procephalothrix* (Iwata, 1960a); on the heteronemerteans *Lineus* (Barrois, 1877; Hubrecht, 1886; Arnold, 1898; Nusbaum and Oxner, 1913; Schmidt, 1934; Iwata, 1957a, 1960a; Gontcharoff, 1960), *Cerebratulus* (Coe, 1899a,c; Wilson, 1900; Kostanecki, 1902a), and *Micrura* (Coe, 1899a; Iwata, 1958, 1960a); on the hoplonemerteans *Drepanophorus* and *Tetrastemma* (Lebedinsky, 1896, 1897), *Geonemertes* (Dendy, 1893; Coe, 1904; Hickman, 1963), *Prosorhochmus* (Salensky, 1884, 1886, 1909, 1912, 1914), *Emplectonema* (Delsman, 1915; Iwata, 1960a), *Prostoma* (Child, 1901; Reinhardt, 1941), *Oerstedia* (Iwata, 1960a), *Carcinonemertes* (Coe, 1902b; Humes, 1942) and *Amphiporus* (Barrois, 1877); and on the bdellonemertean *Malacobdella* (Hoffmann, 1877; Hammarsten, 1918).

In nemerteans two distinct methods of development are recognised. The direct type, occurring in the Palaeo-, Hoplo- and Bdellonemertea, does not involve an intermediate stage and the embryo grows, via a rhabdocoel-like larva, straight into a miniature worm without metamorphosing and incorporating most or all of the embryonic ectoderm, whereas the characteristic indirect development of the Heteronemertea involves a distinctive intermediate larval phase, which metamorphoses into the adult form by replacing the embryonic ectoderm through a process of cellular invagination. The larval ectoderm, shed during metamorphosis, is frequently eaten by the emerging young (Cantell, 1966a,b, 1969). Apart from the direct form, three other larval types, found in heteronemerteans, are the pilidium, very similar to the annelid trochophore, and Iwata larvae which are pelagic, and the Desor larva which is not free-swimming and remains within the egg membrane during its development. The pilidium and direct larvae feed during their larval phases, but the Iwata and Desor larvae do not.

The establishment of a pilidium *auriculatum* in the life cycle of *Hubrechtella dubia* (Cantell, 1969) is the only record of a pilidium larva occurring outside the Heteronemertea. It is not known for certain whether this indicates that the pilidium represents a primitive nemertean larval form, or that *Hubrechtella*, although currently classified as a palaeonemertean, is in fact a heteronemertean possess-

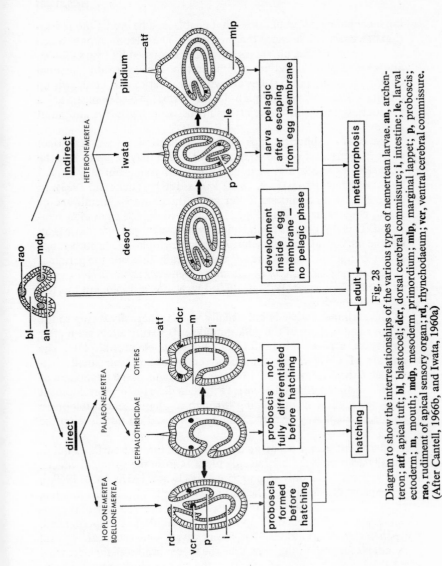

Fig. 28

Diagram to show the interrelationships of the various types of nemertean larvae. **an**, archenteron; **atf**, apical tuft; **bl**, blastocoel; **dcr**, dorsal cerebral commissure; **i**, intestine; **le**, larval ectoderm; **m**, mouth; **mdp**, mesoderm primordium; **mlp**, marginal lappet; **p**, proboscis; **rao**, rudiment of apical sensory organ; **rd**, rhynchodaeum; **vcr**, ventral cerebral commissure. (After Cantell, 1966b, and Iwata, 1960a)

ing morphological peculiarities that enable it to be used as an indication of a close link between the two orders. The relationships between the various postgastrula stages have been discussed by Iwata (1960a) and Cantell (1969), and are summarised in Fig. 28. Iwata (personal communication, Fig. 33) has shown how a series can be drawn up amongst the various anoplan larval forms that demonstrates a gradual alteration in the relationship between the larval and adult axes.

The oval or spherical eggs, before fertilisation, possess a distinct excentric germinal vesicle (egg nucleus) which may be up to half the diameter of the egg, and variable amounts of yolk, externally limited by the egg membrane (Coe, 1895; Iwata, 1960a; Hickman, 1963). In some species an additional outer zone, the glutinous membrane or zona pellucida, is present (Fig. 27A) but is discarded soon after ovulation. This layer swells upon contact with water, and may more than double the overall egg diameter. A conical cytoplasmic protuberance projects from one side of the egg diametrically opposite the germinal vesicle in species such as *Cerebratulus lacteus*, *Lineus torquatus* and *Micrura akkeshiensis* (Wilson, 1900; Iwata, 1957a, 1958), but this too is soon discarded.

The germinal vesicle is not usually visible externally owing to the density and opacity of the yolk globules. Its nature alters soon after ovulation, a polar spindle forming across its equatorial plate. The spindle of *Cerebratulus marginatus* comprises sixteen small ring-shaped chromosomes (Coe, 1899c), and remains in the metaphase of mitosis until fertilisation occurs. Only then does further development take place, division proceeding through to anaphase and telophase (Morse, 1912). Any eggs remaining unfertilised do not develop the first polar body.

Sperm entry can occur at any point on the egg surface, but is usually accomplished on that side furthest from the germinal vesicle. The sperms, commonly 45 μ to 55 μ long (Iwata, 1957a, 1958, 1960a), are capable of penetrating the glutinous as well as the egg membrane, should this still be present. Immediately after entry, a pair of centrosomes and their asters develop in close proximity to the sperm head, as in many invertebrates. The asters divide and a delicate central spindle is formed, at the same time the sperm head rapidly increasing its size by absorbing material from the surrounding egg cytoplasm. The sperm chromatin then becomes rearranged to form a large sperm nucleus, and the first two polar bodies are liberated. Any chromatin remaining at the site of the germinal vesicle forms a fusion nucleus which joins with that of the sperm, the two nuclei at this time being approximately the same size (Coe, 1905a).

The subsequent fate of the centrosomes and sperm asters has not been definitely established; Kostanecki (1902b) believes that the centrosomes persist through to the cleavage stages, but most authors (summarised by Hyman, 1951) regard the cleavage centrosomes as new structures forming after the others have degenerated. The sperm asters clearly do disappear, but their rays may persist beyond the time when distinctive cleavage asters are formed (Coe, 1899c).

The formation of the first and second polar bodies usually takes place within an hour of fertilisation (Iwata, 1958, 1960a), in some cases both then dividing further into two small globules. The significance of this is not at present understood.

Cleavage

Cleavage in nemerteans is of the spiral determinate type, the sequence of development up to the gastrula formation being essentially the same for all species, whether the subsequent pattern followed is direct or indirect. The rate of cleavage varies with the species; in *Cerebratulus leidyi* the first division occurs ten minutes after fertilisation and a definite embryo is formed within nine hours (Coe, 1899c), whereas in *Geonemertes australiensis* cleavage does not commence until the twelve-hour stage and a distinct embryo is not produced until after a few days (Hickman, 1963). The first cleavage plane passes vertically through the animal pole, the second, also vertical, at right angles to the first and forming a four-celled stage comprising approximately equal-sized blastomeres (Figs. 29A–C). Subsequent cleavages, marking the commencement of spiral development, are alternatively dexiotropic and laeotropic (Iwata, 1957a, 1960a). The third, horizontal, division gives rise to the first quartet of animal-pole micromeres (Fig. 29D), which are as large or larger than the macromeres (vegetal pole). Further divisions give rise to other quartets of micromeres (a total of three quartets are formed in *Tubulanus*, four in *Cerebratulus*, *Lineus*, *Malacobdella* and *Prostoma*), then a solid morula of cells (Fig. 29E), and finally a hollow spherical blastula in which the nuclei of the blastomeres enclosing the blastocoel cavity are marginally distributed (Fig. 29F). Gastrulation and endoderm formation usually occurs by embolic, less frequently by epibolic, invagination of the vegetal macromeres (Figs. 29G,H), and involves the fourth quartet of micromeres when this is present. *Prostoma* is somewhat atypical in that endoderm formation is achieved by polar ingression (Fig. 29I) (Reinhardt, 1941). In several species the distinctive cell walls between adjacent blastomeres disappear early on and the origin of the mesoderm cannot therefore be accurately determined. In others the mesodermal tissues clearly arise in one of several ways. It may be derived wholly from the embryonic ectoderm, as in *Malacobdella* (Hammarsten, 1918) and probably *Cephalothrix*, *Oerstedia* and *Procephalothrix* as well (Smith, 1935; Iwata, 1960a), but in *Geonemertes*, *Prosorhochmus* and others certain vegetal ectodermal cells become differentiated into one to four telo-blast cells, formed near the blastopore, which proliferate internally to form endomesodermal tissues (Fig. 29j) that are not arranged into definite bands (Salensky, 1914; Hickman, 1963). The teloblast cells seem to form from cell 4d in most nemerteans, but in *Tubulanus* originate from 3D. In several species micromeric development, possibly of the second quartet, is believed to be the main source of the mesoderm (Hyman, 1951), which is accordingly known as ecto-mesoderm or mesectoderm. In yet others the mesoderm appears to be derived entirely from the irregular ingression of the endodermal components. It seems probable that in several nemertean groups the mesodermal tissues originate via two or more of these routes, but their rapid proliferation and conversion to parenchymatous form prevents further details from being determined.

Direct development

Postgastrula development of the direct type is divisible into two

---

Fig. 29
Embryological development. A–F, Stages in the development of *Geonemertes australiensis*, A, Newly fertilised egg; B, 2-cell stage; C, 4-cell stage; D, 8-cell stage (arrows indicate spiral nature of cleavage); E, Morula; F, Section through blastula; G, 3-day embryo of *Mala-cobdella* showing embolic gastrulation; H, Early embryo of *Cerebra-tulus* showing epibolic gastrulation; I, Embryo of *Prostoma* showing gastrulation by polar ingression; J, Early gastrulation of *Prosorhoch-mus*, showing teloblast cells proliferating endomesodermal tissues; K, 165-hour larva of *Procephalothrix simulus*; L, Sagittal section through 37-hour larva of *Tubulanus punctatus*; M, Sagittal section through 208-hour larva of *Emplectonema gracile*. ar, archenteron; aso, apical sensory organ; atf, apical tuft; b, blastomere; bl, blastocoel; blm, blastocoel filled with mesodermal cells; bp, blastopore; caf, caudal tuft; cm, cir-cular musculature; dcr, dorsal cerebral commissure; e, eye; ecm, ecto-mesoderm; ect, ectoderm; ei, endoderm of future intestine; em, egg membrane; enm, endomesoderm; eno, endoderm; ep, epidermis; gv, germinal vesicle; i, intestine; ima, invaginating macromeres; lm, long-itudinal musculature; ltf, lateral tuft; ma, macromere; mi, micromere; pb, polar body; rd, rhynchodaeum; stb, stylet basis; stm, stomodaeum; tl, teloblast cell; uep, undifferentiated epidermis; vcr, ventral cerebral commissure. (A–F modified from Hickman, 1963; G–J redrawn from Hyman, 1951, after Hammarsten, 1918 (G), after Coe, 1899a (H), after Reinhardt, 1941 (I), and after Salensky, 1909, 1914 (J); K–M re-drawn from Iwata, 1960a)

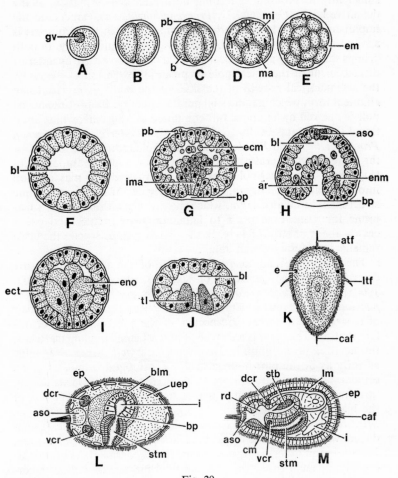

Fig. 29

distinct lines (Iwata, 1960b), the palaeonemertean group (including Iwata's archinemerteans) in which the proboscis does not develop sufficiently to be used in feeding until after the formation of the definitive adult (*Cephalothrix, Procephalothrix, Tubulanus*), and the hoplonemertean–bdellonemertean group, in which the proboscis is advanced enough to be used by the larva for capturing prey. In both groups the radially symmetrical gastrula elongates either by anterior displacement of the apical pole or posterior extension of the body, the asymmetrical growth of the sides of the body giving rise to an elliptical form which assumes bilateral symmetry. In most the apical pole is marked by an apical tuft composed of long cilia, whilst additional shorter caudal tufts are found in *Cephalothrix, Emplectonema, Procephalothrix* and *Tubulanus*, with the cephalothricids also possessing paired dorsolateral cephalic tufts (Smith, 1935; Iwata, 1960a,b) (Fig. 29K). The apical pole sensory plate (Figs. 29G,H) may develop into the adult frontal organ, or may merely be a transient embryonic structure. The ciliated larva, which may rotate within the egg membrane for some time prior to hatching, then escapes and either crawls about or swims feebly. In the ovoviviparous species the larva remains inside the ovarian lining.

The transverse and longitudinal axes of the future worms are determined in the gastrulae of some palaeonemerteans (Iwata, 1960a), since in *Procephalothrix filiformis, P. simulus* and *Tubulanus punctatus* the nervous system rudiments appear at this stage as a pair of enlarged blastomeres, irrespective of the subsequent site of differentiation. The same is also true of at least some hoplonemerteans, the nervous primordium in *Emplectonema gracile* being discernible as early as in the nine-hour gastrula. The foregut originates as an ectodermal invagination (stomodaeum) that grows inwards to join up with the endodermal intestinal rudiments (Fig. 29L). The fate of the blastopore depends upon the species. In some (*Emplectonema, Micrura, Procephalothrix, Tubulanus*), the junction of the stomodaeum with the intestine marks the blastopore position (Iwata, 1958, 1960a), but in others (*Drepanophorus, Prosorhochmus*) the invaginating stomodaeum fuses with the intestinal mesoderm without incorporating the blastopore, which is sealed off (Lebedinsky, 1897; Salensky, 1909, 1914). In the hoplonemerteans in which a separate mouth is missing from the adults, the stomodaeum loses its original opening after invaginating, and connects with the ectodermal rhynchodaeum invagination (Fig. 29M) to form the oesophagus. In the bdellonemerteans (*Malacobdella*) the rhynchodaeal pore is secondarily lost (Hammarsten, 1918), the developing proboscis opening into the dorsal wall of the stomodaeum which then loses its larval mouth and grows forwards to form a new aperture (Figs. 30A–C).

The anus may similarly form as an ectodermal invagination (proctodaeum) that opens into the posterior end of the elongating intestine or, as in *Cephalothrix*, may be formed by the gut growing backwards until it makes contact with and opens at the epidermal surface. Other adult organs originating from the larval ectoderm are the cerebral sense organs, developing as a pair of tubular invaginations that grow inwards towards the cerebral ganglia, the eyes which in most soon sink into the parenchyma, and the epithelial lining of the proboscis, which forms as a tubular ingrowth of the ectoderm just below the frontal glands. The musculature of the proboscis apparatus and body wall, and the general body parenchyma, are derived from the mesodermal tissues of the larva. Mesodermal cells accumulate around the proboscis invagination and then split into two, the layer adjoining the proboscis later delaminating into its muscle layers, the other component similarly giving rise to the lining and musculature of the rhynchocoel. The cavity between the two regions, forming the rhynchocoel, is therefore intramesenchymal and corresponds to Hyman's (1951) definition of a schizocoelom. In the hoplonemerteans the proboscis forms at first as an undifferentiated structure, but as it elongates the three regions characteristic of the order are formed.

The blood system develops as a series of spaces hollowed out in the parenchymatous mesoderm, these gradually fusing to form the pattern representative of the species concerned. The origins of nephridial development have not been definitely established for the palaeo- and hoplonemertean lines of formation, but are generally believed to be ectodermal. The derivation of the various body struc-

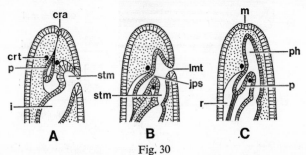

Fig. 30

Formation of the definitive oral aperture and pharynx in *Malacobdella grossa*. **cra**, closed rhynchodaeal aperture; **crt**, cerebral ganglia tissue; **i**, intestine; **jps**, junction of proboscis with larval stomodaeum; **lmt**, larval mouth; **m**, mouth; **p**, proboscis; **ph**, pharynx; **r**, rhynchocoel; **stm**, stomodaeum. (All redrawn from Hyman, 1951, after Hammarsten, 1918)

tures in the direct type of development is shown schematically in Fig. 31.

Indirect development

Indirect development, with one exception (*Hubrechtella dubia*), is confined entirely to the heteronemerteans, and the gastrula grows into an intermediate larva that is either retained within the egg membrane and is never pelagic (Desor larva) or escapes and undergoes a free-swimming phase (Iwata larva and pilidium). The subsequent development of all three is essentially the same.

The Desor larva (Fig. 32A), named after its discoverer (Desor, 1848), is an oval ciliated form without an apical tuft or sensory plate, that does not develop the oral lobes and ciliated oral band found in the pilidium. It characteristically occurs in *Lineus ruber* (Nusbaum and Oxner, 1913).

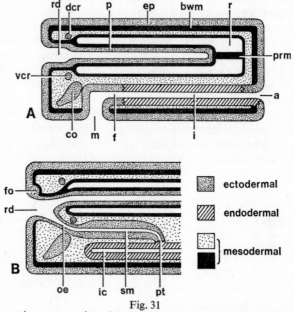

Fig. 31

Schematic representation of the embryological derivation of the principal body structures. a, anus; bwm, body-wall musculature; co, cerebral organ; dcr, dorsal cerebral commissure; ep, epidermis; f, foregut; fo, frontal organ; i, intestine; ic, intestinal caecum; m, mouth; oe, oesophagus; p, proboscis; prm, proboscis retractor muscle; pt, pyloric tube; r, rhynchocoel; rd, rhynchodaeum; sm, stomach; vcr, ventral cerebral commissure.

The Iwata larva (Fig. 32B), so far known only from *Micrura akkeshiensis* (Iwata, 1958), represents an intermediate stage between the Desor and pilidium larvae, although in many respects resembling the larval forms found in direct development (Fig. 28). Iwata's larva can be regarded as a swimming modification of the Desor type, with the formation of an apical tuft and organ, but without the oral lobes and ciliation of the pilidium. It does, however, resemble the pilidium in possessing a larval ectodermal layer. Neither the Desor nor Iwata larvae feed during their development.

The pilidium larva (Fig. 32C), first discovered and named by J. Müller (1847), is typically a rather helmet-shaped organism with a thin ciliated larval ectoderm and a gelatinous and transparent appearance. A pair of lateral lobes, one either side of the mouth, extend downwards and flank the oral aperture. At the apical surface an invaginated sensory plate, composed of attenuated columnar cells, bears one or more tufts of long cilia. The pilidium larva is known from several heteronemertean genera.

Several authors, including Bürger (1895), Schmidt (1931a, 1937) and Dawydoff (1940), have devised classification schemes for pilidia based principally upon their external morphology, but such systems are necessarily artificial since in many cases the larvae have not been related to adult nemerteans. Cantell (1966b, 1969), in recent studies of pilidium development, includes examples from the groups designated as *gyrans* (*Lineus albocinctus*, *L. bilineatus*), *pyramidale* (*Micrura purpurea*) and *recurvatum* (Baseodiscidae?), as well as *recurvum* and *incurvatum* of unspecified adults (Figs. $32c_1-c_3$).

The larvae possess a densely ciliated sac-like intestine with no anus, opening via a narrow buccal ridge into a cavernous foregut and mouth. The mouth and foregut are of ectodermal origin, the intestine endodermal, and their junction corresponds to the blastopore. Feeding is accomplished by a ciliary mechanism (see Chapter 2).

In several pilidia pigmented chromatophores are arranged around the oral margins of the body, particularly in the anterior and lateral lobes (Cantell, 1969). The margins of the lobes in some forms also bear denticle-like outgrowths, the number increasing with age, but their significance is not understood.

The blastocoel of the larvae is filled with a viscous fluid containing irregularly dispersed amoeboid entomesodermal cells, formed from mesoblasts derived originally, at least in some species, from cell 4d. As the larvae grow, several of these blastocoel cells become attached by their processes, and differentiate into muscle cells that together comprise the complex musculature of the body, extending from the apical plate to the oral margins (Fig. 32C).

Metamorphosis

Metamorphosis in all three larval types is accomplished by a series of ectodermal or amniotic invaginations, first discovered by Metschnikoff (1869). There are five in the Iwata larva (an unpaired posterior dorsal and a pair each of anterior cephalic and posteroventral trunk invaginations), seven in the pilidium (with the addition of a pair of lateral cerebral invaginations), and eight in the Desor larva (an extra unpaired anterior proboscis invagination). The paired amniotic areas become cut off from the larval ectoderm as small compressed sacs (Fig. 32D), each enclosing an amniotic cavity, with thin outer (amnion) and thick inner (blastodisc) walls (Fig. 32E) (Iwata, 1958), whereas the unpaired regions, except in Iwata's larva, are formed by delamination without amnion production.

The blastodiscs are centres of active cell division, and rapidly enlarge and spread radially, flattening out as they do so. Eventually their boundaries meet and merge, their complete fusion giving rise to the definitive epidermis which encloses the gut and mesodermal tissues. Around the periphery the fused amnia form a thin embryonic membrane.

By this time the anteroposterior axis of the future nemertean has become established. In the Desor larva the axis of the metamorphosing worm is the same as that of the original larva, but in the Iwata form it is opposite (Fig. 33K), in the pilidium at right or oblique angles (Figs. 33E–J), to that of the larval envelope.

Fig. 32

Larval forms. **A**, Sagittal section of advanced Desor larva of *Lineus*; **B**, Sagittal section of Iwata larva of *Micrura akkeshiensis*; **C**, Sagittal section of advanced pilidium larva of *Cerebratulus*; C_1–C_3, Pilidium types; *pilidium recurvatum* (C_1), *pilidium recurvum* (C_2), and *pilidium incurvatum* (C_3); **D**, Sagittal section of 53-hour Iwata larva showing early metamorphosis; **E**, Amnion formation and disc invagination in Desor larva of *Lineus*; **F**, Advanced pilidium of *Cerebratulus* containing fully metamorphosed miniature worm. **am**, amnion; **amc**, amniotic cavity; **aso**, apical sensory organ; **atf**, apical tuft; **blm**, blastocoel filled with mesodermal cells; **bp**, blastopore; **cb**, ciliated band; **cbs**, cephalic blastodisc; **ch**, chromatophore; **cob**, cerebral organ blastodisc; **dbs**, dorsal blastodisc; **e**, eye; **enm**, endomesoderm; **ep**, epidermis; **f**, foregut; **i**, intestine; **la**, larval aperture; **le**, larval ectoderm; **lmt**, larval mouth; **mcl**, muscle cell; **mf**, muscle fibre; **mlp**, marginal lappet; **mlv**, metamorphosed larva; **p**, proboscis; **pi**, proboscis invagination; **rp**, rudiment of proboscis; **std**, stomodaeal diverticulum; **stm**, stomodaeum; **tbs**, trunk blastodisc. (A redrawn from Hyman, 1951, after Arnold, 1898; B, D redrawn from Iwata, 1958; C after Hyman, 1951; C_1–C_3 redrawn from Cantell, 1966b; E redrawn from Hyman, 1951, after Schmidt, 1934; F modified from Hyman, 1951, after Verrill, 1892)

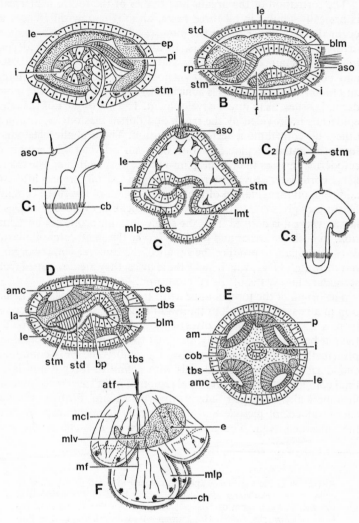

Fig. 32

The formation of the organs and tissues of the future nemerteans is basically similar in all three cases, although some differences do occur that can be related to the number of ectodermal invaginations involved in the metamorphosis. In all, the embryonic endodermal intestine is retained by the adults, and the formation of the proboscis and body-wall musculature, rhynchocoel and rhynchocoelic lining, and blood vascular system is achieved by differentiation of the meso-dermal tissues, as in direct development. The major part of the adult epidermis is provided by the trunk and dorsal blastodiscs, although others do contribute a certain proportion. The cephalic blastodiscs form the anterior epidermis, but also bud off internally a pair of ectodermal masses that develop into the cerebral ganglia. The lateral nerve cords later grow posteriorly from the ventral cerebral lobes. In addition, an invagination of the median region of the fused cephalic blastodiscs gives rise to the proboscis epithelium, except in the Desor larva, where it is derived from the single anterior proboscis invagina-tion. In both the Desor and pilidium larvae the small cerebral blasto-discs invaginate to produce the canals of the cerebral sensory organs, but in Iwata's larva, which lacks these discs, they are developed from secondary ingrowths of the stomodaeum.

The origin of the foregut appears to be via a stomodaeal invagina-tion in all cases, although in the Iwata larva it is secondarily derived from a dorsal stomodaeal diverticulum. During later development the gut becomes considerably modified to attain its adult condition, principally by the replacement of the embryonic yolk-containing endodermal and stomodaeal cells with definitive foregut and intes-tinal components. In the intestinal region of most species the pseudo-metamerically arranged lateral diverticula are formed through encroachment of mesodermal tissues rather than by outgrowths of the intestinal wall. The anus originates as a small proctodaeal

---

Fig. 33

Alteration in the axiality of anoplan nemertean larval forms. **A**, Gas-trula of *Procephalothrix simulus*; **B**, Young larva of *Procephalothrix simulus*; **C**, Later larva of *Procephalothrix simulus*; **D**, Larva of *Tubu-lanus punctatus*; **E**, *pilidium incurvatum* type (adult unknown); **F**, Pili-dium of *Hubrechtella dubia* (*pilidium auriculatum* type); **G**, Pilidium of *Lineus albocinctus* (*pilidium gyrans* type); **H**, Pilidium of *Micrura purpurea* (*pilidium magnum* type); **I**, Pilidium of Baseodiscidae (*pilidium pyramidale* type); **J**, Pilidium variant of Baseodiscidae (*pilidium mag-num* type); **K**, Iwata larva of *Micrura akkeshiensis*; **L**, Desor embryo of *Lineus ruber*. A–F, Palaeonemertea (A–C = Archinemertea of Iwata, 1960a), G–L, Heteronemertea. →, adult axis, arrowhead anterior; —o, larval axis. (From an original drawing by Dr Fumio Iwata)

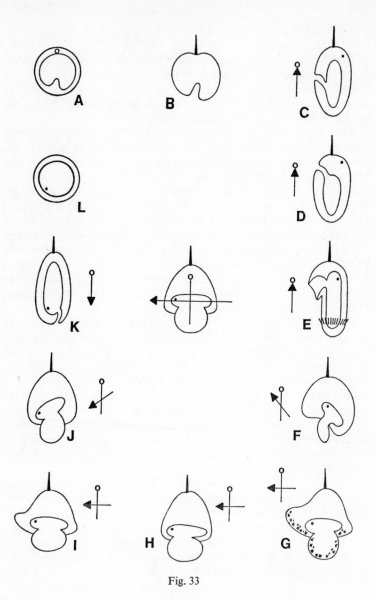

Fig. 33

(ectodermal) ingrowth that joins up with the posterior intestine, as in some direct examples.

The origins of the nephridial system are not definitely established for species developing via Desor or pilidia larvae, although variously identified as ectodermal or stomodaeal, but in the Iwata larva of *Micrura akkeshiensis* are almost certainly derived from splitting up of the ventral surface of the stomodaeum.

After the completion of metamorphosis, the young nemertean remains for only a short time within its larval case (Fig. 32F). The larval ectoderm, including the apical sense organ of the pilidium and Iwata larva, and amnion are then discarded, and the miniature adult emerges to commence its normal free-living existence. The Iwata and Desor larvae do not apparently utilise their larval envelope as a food source at this time (Nusbaum and Oxner, 1913; Iwata, 1958), but in the majority of cases the metamorphosed pilidium devours part or all of the larval tissues during hatching (Cantell, 1966a, 1969). The principal exception to this general rule is *Hubrechtella dubia*, which leaves its case intact (Cantell, 1969), possibly since in this species the young nemertean is rather more loosely attached to the larval tissues than are the heteronemertean forms.

Hyman (1951) states that the Desor type of larva appears to be derived from the pilidium grade as an adaptation to the more variable conditions encountered in shallower waters. Whether or not this interpretation is correct is difficult to ascertain, but there are certainly far more littoral and coastal heteronemertean species with pilidia larvae than with the Desor type in their life cycles. It could equally well be argued that the Desor larva is the more primitive of the heteronemertean larval types, and that the pilidium is a form evolved to ensure a wider dispersal of the species. That the Iwata larva is an intermediate grade between the other two does not benefit either hypothesis.

How long it takes for the young nemerteans to attain sexual maturity is not known, but it appears that the establishment of the sexual characteristics occurs early on in postembryonic development. In *Lineus ruber*, a strictly dioecious species, the sexual development of young, immature or regenerating specimens is strongly inhibited by a hormonal gonado-inhibiting factor (F.G.I.) secreted by the cerebral ganglia (Bierne, 1970a,b,c). In addition sexual differentiation is apparently at least partly regulated by a second, androgenic, hormonal substance (F.A.) produced by the males, since in experimental studies on heterosexual chimeras sex reversal of the ovaries is regularly caused. Bierne's (1970a) 'strong suggestion' that reproduction is governed by an endocrine system seems well justified.

Experimental embryology

Several workers have investigated development in the eggs of nemerteans in attempts to ascertain the stage at which determination occurs, particularly in *Cerebratulus* species (Wilson, 1903; Zeleny, 1904; Yatsu, 1904, 1910a,b; Hörstadius, 1937).

Yatsu (1904, 1910b) obtained various types of defective larvae by removing cytoplasm from eggs at different developmental stages between pre-ovulation and first cleavage, the incidence of abnormality increasing the nearer to cleavage the experiment was carried out. Determination appears to be initiated by the dissolution of the germinal vesicle, but mature eggs used from before this stage also give rise to a certain, but smaller, number of atypical pilidia. Egg fragments taken at the metaphase of the first maturation (polar) mitosis subsequently develop abnormalities particularly in the apical organ, ciliated lobes and gut, and it seems that the animal and vegetal polar regions of the egg differentiate independently. Removal of tissue from the lower, vegetal pole at the two-cell stage has no apparent effect upon the size relationships of the eight-cell stage blastomeres, whereas some variation can be induced by cutting off a portion of the animal pole (Yatsu, 1910a). Any defects caused by cytoplasmic excision are not replaced during later larval growth; they do, in fact, become more obvious as development proceeds. There are, therefore, no compensatory mechanisms within the growing pilidium whereby lost components can be refurbished.

If, on the other hand, egg fragments are taken before the germinal vesicle is fully formed, i.e., during the ripening stages, the subsequent cleavage and development is perfectly normal. These results indicate that determination is effected at some time soon after egg maturation but well before fertilisation, although being reinforced by sperm entry and fusion of the egg and sperm nuclei. Determination can only proceed to a certain level in the absence of sperm entry, since eggs with cytoplasm removed at the first mitotic metaphase and then fertilised at different time intervals show approximately the same incidence of defective pilidia. If the process of determination continued after metaphase in the absence of sperm entry, the abundance of abnormal larvae would be greater after a five-hour interval between metaphase and fertilisation than after one hour. Since ripe eggs are found within the ovaries, determination to at least some degree must be present even before ovulation.

The factors controlling morphogenesis do not appear to be the same as those regulating cleavage. The mode of division is not disturbed by cytoplasmic excision unless performed very close to the onset of cleavage, and damage without material removal has little or no effect even if occurring during division (Yatsu, 1910a).

Blastomeres isolated at the two- or four-cell stages develop quite normally, although the pilidia subsequently formed are considerably smaller than usual. After the eight-cell stage, however, isolation of individual cells or equatorial splitting of the embryo results in the loss of particular larval components. The first quartet of micromeres (animal pole) will, if isolated from the macromeres, give rise to a small larva with a more or less normal apical sense organ, but the ciliated lobes and archenteron are either totally lost or severely reduced. This condition Yatsu (1910b) calls an anenterion. Conversely, isolated macromeres may fail to form an apical organ but develop an unusually large archenteron (Zeleny, 1904; Hörstadius, 1937). Determination is therefore effected quite precisely at the third cleavage, and no evidence of this can be detected at the four-cell stage.

The manner whereby cleavage takes place can be upset by rearing eggs in virtually calcium-free sea water, or by the application of compression (Yatsu, 1910a,b). The so-called ring- or plate-embryo formed will develop into a pilidium-like larva but will always possess some morphological abnormality. The presence of environmental calcium is of great importance to normal embryological development, and in its absence differentiation and morphogenesis are somewhat inhibited. The effect is, however, reversible, and embryos replaced in normal sea water will rapidly recover their normal differentiative ability.

The eggs of *Cerebratulus lacteus* and *C. marginatus* have also been used by Morse (1912) for experiments on artificial parthenogenesis and hybridisation. Although sperm homogenates do not induce egg development, treatment with certain chemicals, including HCl, KCl, NaOH and saponin, has variable effects but some irregular segmentation can be stimulated. In many cases polar bodies are formed, but rarely does development proceed beyond this stage, and no parthenogenetic embryos could be induced to grow further than the early blastula condition.

Experimental hybridisation with molluscan eggs or sperm were even less successful, although some 'embryos' did start irregular cleavage, the sperm donor being *Ilyanassa obsoleta*.

6

ECOLOGY AND ZOOGEOGRAPHY

Geographical distribution

Nemerteans are found all over the world, in every ocean and larger sea extending from the Arctic to the Antarctic. They are predominantly marine organisms, most species occurring in littoral or coastal regions, but a few examples have successfully colonised brackish-water, freshwater or terrestrial habitats, and the pelagic polystyliferous hoplonemerteans characteristically live in the abyssal oceanic depths.

Apart from the true pelagic species, which either float passively or swim sluggishly, nemerteans are benthic in habit, living beneath embedded boulders and rocks, amongst the holdfasts or fronds of algae, in narrow crevices and coral reefs, with various colonial sessile invertebrates (e.g., bryozoans, barnacles and mussels), or burrowed into sands, muds or gravels. A few forms live in tubes, either lining them with their own viscous mucus (*Hubrechtella*), or taking over the empty burrows of such polychaetes as *Chaetopterus* or *Pomatoceros* (*Lineus*, *Micrura*, *Tubulanus*) or amphipods as *Haploops* (*Amphiporus*) (Brunberg, 1964; Kirsteuer, 1967a). Few of the nemertean genera have evolved parasitic representatives, and in most cases the association appears to be one of commensalism rather than strict parasitism.

The problems inherent in nemertean taxonomy, with many forms so incompletely described that later findings cannot with certainty be allocated to particular species, have meant that for the vast majority we have comparatively little idea as to their distributional limits. Since the bulk of the world's land masses occur in the northern hemisphere, it is only natural that the nemertean fauna should be better known from this rather than from the southern global regions.

Hyman (1951) states that the littoral species are found in the greatest variety along the northern European and Mediterranean coasts, but this observation can no longer be regarded as justified in the light of more recent investigations. Brazil, for example, possesses a particularly rich nemertean fauna, and both North American coasts harbour a wide range of both genera and species. In general, however, temperate and subpolar regions possess both greater numbers and varieties of nemerteans than do tropical or subtropical areas.

Within the northern hemisphere several species exhibit circumpolar ranges; amongst these the palaeonemertean *Cephalothrix linearis*, the heteronemerteans *Cerebratulus marginatus*, *Lineus bilineatus*, *L. geniculatus* and *L. ruber*, and the hoplonemerteans *Emplectonema gracile*, *Oerstedia dorsalis* and *Tetrastemma candidum*, extend from Japan, across the northern Pacific and Atlantic, and into the Mediterranean and North Sea (Coe, 1943; Iwata, 1952, 1954a; Gontcharoff, 1955; Riedl, 1959; Kirsteuer, 1963a; McCaul, 1963; Uchida *et al.*, 1963; Brunberg, 1964; Corrêa, 1964). Other species, the boreal forms, do not apparently have such an extensive distribution, although future records may well show that several of these are in fact circumpolar in their range. *Lineus torquatus* and *Paranemertes peregrina*, for example, occur both in Japan and down the Pacific seaboard of North America, but have not been recorded from the Atlantic.

Most nemertean species have been recorded within far narrower geographical limits, and often from only single localities. Many of these apparently indigenous forms will almost certainly possess wider ecological ranges than are at present known, but undoubtedly several will be restricted in their distribution, even though their limiting factors are not known. Southern hemisphere and equatorial nemerteans appear to fall mainly into this zoogeographical category; many of the Brazilian or Chilean species, for example, are not known other than from the shores of these countries (Corrêa, 1954, 1957, 1958; Friedrich, 1970). On the other hand some of the northern species do extend across the equator into the southern hemisphere. *Lineus ruber*, *Tubulanus nothus* and *Tetrastemma candidum* are species that have been found from localities as far apart as Alaska and South Africa (Corrêa, 1964), as well as from many intermediate places. Several of the bathypelagic hoplonemerteans are also known from both sides of the equator, particularly wide ranges being shown by *Nectonemertes pelagica*, *N. minima* and *Pelagonemertes rollestoni* (Coe, 1926).

One of the major problems concerning geographical distribution is that of variation within a species. The wide range of morphological variation that can occur within a single species is exemplified by the work of Hickman (1963) in his study of the terrestrial hoplonemertean *Geonemertes australiensis*. Several of the currently recognised littoral species that are obviously closely related may, with further study, prove to be variants of the same form and their geographical limits accordingly extended. Coe (1943) lists several examples of this taxonomic dilemma which clearly illustrate the difficulties of positive identification. *Oerstedia dorsalis*, a small but common species, is perhaps the most variable of nemerteans from the point of view of colour and may be spotted, banded, mottled or striped in red, brown, green or yellow. Conversely, six supposedly distinct species of *Amphiporus* (*arcticus, heterosorus, multisorus, stimpsoni, superbus* and *thompsoni*) may well simply be variants of *Amphiporus angulatus*, a widely distributed form from which they are distinguished only by minor differences in colour or eye arrangement.

Littoral and coastal nemerteans

Many more nemertean species occur in littoral and coastal habitats than in all other types of niche combined. They extend from the high-tide level downwards, those of the upper littoral reaches living in essentially semi-terrestrial conditions. Such types as *Acteonemertes bathamae* (Pantin, 1961a), *Prosorhochmus claparedii* (Pantin, 1969) and *Geonemertes nightingaleensis* (Brinkmann, 1947) form transitional species living between littoral and terrestrial conditions, frequently extending their range beyond the supralittoral fringe and above the splash zone.

Most littoral species, however, cannot survive beyond the upper tidal limits, and are usually found in the mid- or lower-shore reaches. The number of species present increases as the low-tide level is approached, although many are both particularly small and easily overlooked.

The nemertean fauna of rocky shores is generally far richer than that of other intertidal regions, many species being commonly found either beneath stones partly embedded in coarse or fine gravels and muddy sands, or amongst the holdfasts of fucoids and particularly *Laminaria*. Most of the littoral nemerteans show a marked dependence upon salinity and are therefore classed as stenohaline organisms. On open oceanic coastlines, especially under exposed conditions, such forms as *Emplectonema gracile* and *Lineus bilineatus* may be found well up the shore, but in enclosed and sheltered areas subject to considerable salinity dilution (e.g., the Kattegat and lower Baltic) occur only in the sublittoral regions and do not extend above the lower tidal limits (Brunberg, 1964).

F*

Many of the littoral species normally have the lower boundaries of their ecological range extending into the subtidal coastal waters (Corrêa, 1964), and a few benthic forms, such as *Carinina grata*, are known from depths as great as 2500 metres or more (Coe, 1943). The range of nemertean types encountered on sandy and muddy shores is usually considerably smaller than that from rocky regions, although in sheltered muddy bays containing dense beds of eel-grass (*Zostera*) nemerteans may be particularly abundant (McCaul, 1963; Kirsteuer, 1963a,b). Several types, such as *Lineus bilineatus* and *Micrura fasciolata*, may be found on muddy and sandy as well as rocky shores, but more often nemerteans inhabiting muds and sands are characteristically confined to these types of substrates. Various species of *Cerebratulus*, particularly the large *Cerebratulus marginatus*, form typical examples, but others found include *Amphiporus*, *Hubrechtella*, *Lineus*, *Paranemertes*, *Poseidonemertes*, *Tubulanus* and *Zygeupolia* (Coe, 1944a; Iwata, 1952; McCaul, 1963; Uchida *et al.*, 1963; Brunberg, 1964; Kirsteuer, 1967a). Many of the psammo-biontic (sand-dwelling) species are small, ranging from one to ten millimetres in length (Kirsteuer, 1967b), and accordingly the geographical distribution of such nemerteans is perhaps the least well known. Species of *Ototyphlonemertes*, for example, although well known from European shores by the beginning of the twentieth century (Bürger, 1897–1907), were not recorded from South America until several decades later (Corrêa, 1948).

The littoral nemerteans are not as a rule active either during the day or when uncovered by the receding tide, but two mud- or sand-dwelling species, *Paranemertes peregrina* (Roe, 1971) and *Tubulanus annulatus* (Gibson, unpublished), are actively predatory during the low-tide periods, moving about in search of food on the damp surface or in calm shallow water.

Very few nemerteans have had their ecology investigated in detail, and exceptionally little is known about the group as a whole. Kirsteuer (1963a,b), in a study of *Oerstedia*, *Oerstediella* and *Tetrastemma* species from Adriatic and other European waters, found that some possessed quite restricted ecological limits, whereas others occurred in a wide variety of habitats. Similar conclusions, but with fewer species, had earlier been reached by Riedl (1959).

From a total of 961 nemerteans recovered, nearly 50 per cent belonged to these three genera. Their relative abundance and occurrence in the various types of localities is shown in Table 4.

Two points immediately obvious from the Table are that more species and individuals are found in the plant communities than from ground habitats, and that in all the situations in which it occurs

Table 4
The distribution of certain hoplonemertean species in six types of
habitat from the Adriatic (from Kirsteuer, 1963a).

Nemertean species	Plant habitats			Ground habitats		
	Cystosira	Zostera	Vidalia + Udotea	Below rocks from Zostera beds	In ground beneath rock level	Mud
T. fulvum	0·3					
T. vastum	1·4					
T. flavidum	3·2					
O. tenuicollis	1·1	1·0				
Oe. dorsalis	0·8	0·3	0·1			
T. vittigerum	0·3		1·3			
T. coronatum	1·8	1·0	8·2	0·3		
T. melanocephalum	3·2		0·6	0·3		
T. candidum	9·8	2·0		2·3	10·0	
T. helvolum	3·4		0·1		0·6	
T. peltatum				0·8	0·4	
T. diadema				2·7	2·2	
T. longissimum			2·6			
T. vermiculus			0·1		1·2	
T. virgatum						3·2

Figures indicate the relative abundance of each species: O = Oerstediella
Oe = Oerstedia   T = Tetrastemma

Tetrastemma candidum is either the dominant or a major representative. In the Vidalia : Udotea community it is apparently replaced by
Tetrastemma coronatum. Most of the other species, as judged from
their distribution patterns, show distinct preferences for particular
habitats and it can be inferred from this that intraspecific competition
is one contributory factor regulating the numbers found of each. A
detailed investigation especially of Cystosira communities might
yield interesting results.

In his 1963b paper Kirsteuer has considered the ecology of nemerteans from a systematic aspect, and there appears to be a quite
definite relationship between the habitat occupied and the grade of
nemertean organisation. With the exception of the bdellonemerteans,
which are exclusively commensals of molluscs, each order tends to be
more characteristically associated with one type of substrate or
habitat than another, although there is no absolute correlation and
some geographical variation can be found. In general, however,
palaeonemerteans are either absent or comprise only a minority of
the nemertean fauna in many plant communities, whereas in these
surroundings hoplonemerteans form the dominant group. Con-

versely, the palaeonemerteans can form the principal nemertean inhabitants of muddy shores, although they do not do so invariably, whereas hoplonemerteans, even when present in conspicuous numbers, are never the major form. Heteronemerteans occur in all types of situations, but tend to be more abundant in sands and muds although forming a large proportion of the nemertean numbers in certain plant communities. The species of *Oerstedia*, *Oerstediella* and *Tetrastemma* listed in Table 4 together comprise from 34 per cent of all nemerteans found in the mixed *Vidalia* : *Udotea* community up to 100 per cent of the *Zostera* forms, and from 58 per cent of the bottom forms within the *Zostera* beds to only 16 per cent of the nemertean inhabitants of muds.

Kirsteuer's findings are entirely supported by the work of Brunberg (1964), who recorded the distribution of nemerteans from Danish waters according to their substrates. Her results are indicated in Table 5.

Little quantitative information is available on the density of nemerteans, although many species are recorded variously as 'abundant' or 'common'. Several species exhibit gregarious habits, either throughout the year or, more often, in the breeding season, and clearly these forms will possess a high but localised density. The author has found 'balls' of twenty or thirty *Cephalothrix* knotted together under stones on the Yorkshire coast, although from one mass to another may involve distances of many metres. Amongst those nemerteans that are not normally gregarious, Roe (1971) found that the mud-dwelling *Paranemertes peregrina* showed a maximum average density of about fourteen worms per metre square, although this number varies greatly depending upon the time of the year, and was mostly far less than this. Leppäkoski (1969) found only five *Micrura* per metre square from the Borholm Basin in the southern Baltic.

In a recent study of the distribution of *Lineus ruber* around the shores of Anglesey, North Wales, preliminary results suggest that, although the nemerteans are found on a wide variety of beaches and extend from the upper littoral regions downwards, they possess a quite narrow range of habitat preferences. This work is currently in progress, but sufficient evidence is available to indicate that the occurrence of *Lineus ruber* is regulated by a combination of substrate selection, density of potential predators, angle of slope of the shore, and height in relation to tides (D. G. Eason, personal communication).

The collection of nemerteans requires somewhat special care since many forms are both small and extremely fragile. A most useful

Table 5
The distribution of Danish nemerteans according to substrate type.

Genus	Polluted muds	Haploops colonies in muds	Amphiura colonies in muds	Sand	Shell gravel	Stones	Laminaria, Fucus, etc.	Ulva, Zostera
Palaeonemerteans								
Carinina (2)		+	++					
Tubulanus (4)	+	++	+	(+)(+)	+	++	(+)	
Cephalothrix (2)								++
Callinera (1)	+	+		(+)				
Hubrechtella (1)	+	+	+					
Heteronemerteans								
Lineus (3)	+	++	+	(+)+	+	(+)+	(+)+	++
Micrura (2)		++	+	+	+	++	(+)+	
Cerebratulus (4)?	+		+					
Hoplonemerteans								
Emplectonema (1)							+	
Nemertopsis (1)	+	+						
Amphiporus (4)		+++	(+)(+)	+	+	(+)+	+	+
Tetrastemma (2)	+			++			+	+
Prostomatella (2)				+			+	
Oerstedia (1)		(+)				(+)	+	
Oerstediella (1)							+	

Bracketed numbers indicate the number of species in each genus. Each cross represents the occurrence of a particular species (i.e., two of the four *Tubulanus* species were found under stones); crosses inside brackets indicate only occasional recoveries. Modified from Brunberg (1964).

paper dealing with effective methods for extracting nemerteans from different substrates is that of Kirsteuer (1967b). Just how essential it is to use the correct procedure in collecting nemerteans is clearly illustrated by Kirsteuer's observation that 'a sample of *Cystoseira barbata* (brown algae), which was collected from a boat and brought in from a depth of 2·5 metres with a rake, contained only 7 per cent of the nemertean fauna that was obtained from a similar sample taken from the same place by a diver'.

A selection of useful articles listing the known species of littoral and coastal nemerteans from various world localities is given below:

Antarctica: Bürger, 1904a; Coe, 1950; Southgate, 1957.
Arctica: Bürger, 1904b; Coe, 1952; Friedrich, 1957.
Atlantic North America: Coe, 1943; McCaul, 1963.
Baltic Sea: Friedrich, 1935, 1936.
Black Sea: Müller, 1962, 1966.
Brazil: Corrêa, 1948, 1954, 1957, 1958, 1966.
British Isles: McIntosh, 1873–4.
Chile: Friedrich, 1970.
Curaçao: Stiasny-Wijnhoff, 1925; Corrêa, 1963.
Denmark: Brunberg, 1964.
Falkland Islands: Wheeler, 1934.
France: Gontcharoff, 1955.
Greenland: Coe, 1944b.
Gulf of Mexico: Coe, 1951a,b, 1954; Corrêa, 1961.
Hawaii and Marshall Islands: Coe, 1934c, 1947.
Iceland: Friedrich, 1958, 1960.
Indian Ocean: Punnett and Cooper, 1909; Wheeler, 1937.
Indochina: Dawydoff, 1952.
Japan: Yamaoka, 1940; Iwata, 1951, 1952, 1954a,b,c, 1957b;
    Uchida et al., 1963.
Madagascar: Kirsteuer, 1967a.
Mediterranean: Bürger, 1895; Wijnhoff, 1913; Stiasny-Wijnhoff,
    1926; Friedrich, 1940; Corrêa, 1956; Riedl, 1959; Kirsteuer,
    1963a.
Mozambique Channel: Kirsteuer, 1965.
North America: Coe, 1905b.
North Sea: Punnett, 1903; Wijnhoff, 1913; Friedrich, 1936;
    Remane, 1940.
Pacific Central America: Coe, 1940b.
Pacific West and Northwest America: Coe, 1901, 1905a, 1944c;
    Corrêa, 1964.
Porto Rico: Coe, 1902c.
Russia: Ouchakoff, 1925; Yashnov, 1948.
South Atlantic: Wheeler, 1934.
South and West Africa: Wheeler, 1934, 1940; Stiasny-Wijnhoff,
    1942.
Sweden: Hylbom, 1957.
Virgin Islands: Corrêa, 1961.
General Zoogeography: Hubrecht, 1887a; Bürger, 1897–1907.

Estuarine and brackish-water nemerteans
Although several nemertean species are known to penetrate estuarine
and brackish-water habitats, comparatively few appear to be speci-

fically adapted to live in conditions of reduced salinity. McCaul (1963), for example, records that of twenty-two species found on Virginian shores, most could also be recovered from estuarine regions, including such forms as *Carinoma tremaphoros*, *Tetrastemma vermiculus*, *Tubulanus pellucidus* and *Zygeupolia rubens*, all of which have a normal distribution extending several metres below low-tide level.

In theory any of the upper- and mid-shore littoral species are potentially capable of living at least on the fringes of estuarine conditions, since the very nature of the intertidal environment is indicative of regular exposure to reductions in salinity. Brunberg (1964), however, notes that only a few of the marine nemerteans occurring in Danish waters do not show a marked dependence upon salinity (*Amphiporus lactifloreus*, *Lineus ruber*, *Malacobdella grossa* and *Tetrastemma melanocephalum*), most being intolerant of dilution. Two hoplonemertean species that occur in brackish rather than marine conditions are *Amphiporus cordiceps* and *Prostomatella obscura*, the latter being particularly abundant in the Gulf of Finland in salinities of 6‰ or less.

Iwata (1970) describes three nemertean species from the brackish waters of Lake Hinuma, Central Japan, all with very limited distributions. The salinity of the lake varies from 60 per cent sea water in summer to as little as 15 per cent in winter, compared with the open sea. These species, *Hinumanemertes kikuchii* (Heteronemertea) and the hoplonemerteans *Sacconemertella lutulenta* and *Sacconemertopsis olivifera*, are found in mud from the lake bed.

Around the British Isles, Green (1968) lists *Lineus ruber* and *Tetrastemma coronatum* as common inhabitants of estuarine sands and muds, *Lineus ruber* in addition being frequently found in salt-marsh pools. The lower limit of tolerance for *Lineus* is about 8‰ salinity according to Remane (1958), although D. G. Eason (personal communication) has found that this species will survive several hours' immersion in water as diluted as 3·5–4‰.

Iwata (1970) has found that the Lake Hinuma species, which are permanently exposed to lower salinities, possess a different development of their blood vascular and nephridial systems compared with those of nemerteans from marine environments. His interpretation of this situation is that the structures are actively involved in the osmoregulatory mechanisms of the body, and represent a morphological adaptation to brackish-water conditions. Such a belief at first appears logical, but has not yet been conclusively demonstrated and is somewhat confounded by the lack of similar development in freshwater forms such as *Prostoma*.

Most nemerteans, if exposed to diluted sea water, absorb fluid and become swollen and turgid; the greater the dilution experienced the more distorted their bodies become. There is at present no conclusive evidence to show whether species like *Lineus ruber*, which can tolerate such maltreatment, are capable of slow osmoregulation or merely able to survive a wide range of body-fluid dilutions. Some authorities report that normal body dimensions are recovered within a comparatively short time, but others indicate that turgidity is retained until the animals are replaced in normal sea water. Marine species intolerant of dilution, and therefore incapable of penetrating estuarine or brackish-water habitats, are presumably either unable to alter their osmoregulatory mechanisms (if they have them) to cope with the increased influx of water, or are simply unable to survive even a temporary dilution of their cellular fluids. This ecologically important aspect of nemertean physiology appears to warrant a good deal of further investigation.

Freshwater nemerteans

Very few freshwater nemertean species have yet been reported, and none from the palaeonemertean group. The distribution of freshwater forms is quite sporadic and, in most instances, species are known from only single localities. The limnetic nemerteans currently listed are:

Heteronemertea: *Planolineus exsul, Siolineus turbidus*
Hoplonemertea: *Campbellonemertes johnsi, Prostoma eilhardi, P. graecense, P. grande, P. jenningsi, P. lumbricoideum, P. padanum, P. puteale, P. rubrum, Potamonemertes percivali*
Bdellonemertea: *Malacobdella auriculae*

The genus *Prostoma* is by far the most widespread of the freshwater forms, particularly the two species *eilhardi* and *graecense*. Both have been recorded from Europe and Africa; *eilhardi* also occurs in Brazil and possibly Argentina and Uruguay, and *graecense* has a range extending to the British Isles, Japan, Australia and Tasmania. The original localities of these species is not known, but their widespread distribution is generally believed to be due primarily through the importation and exportation of freshwater vegetation, secondarily on the feet of water birds. Coe (1959) suggests that these are the routes whereby the North American species (*Prostoma rubrum*) has achieved a wide but intermittent dispersal throughout the United States.

Moore and Gibson (1973) suggest that the limnetic hoplonemerteans have evolved from marine ancestors by two distinct routes.

These can be termed the tetrastemmid (*Prostoma*) and prosorhochmid (*Campbellonemertes, Potamonemertes*) lines, their origins being
postulated from the known habitats of near relatives. The evolutionary sequence, with examples of genera from intermediate habitats, is:

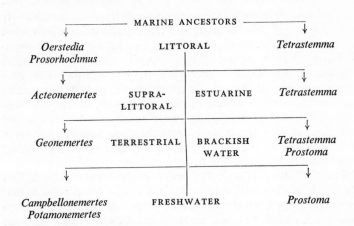

This postulated evolutionary divergence of the two groups is
supported by such evidence as the occurrence of *Prostoma graecense*
(with *Sacconemertes arenosa* and *Tetrastemma obscurum*) in brackish
regions of the Gulf of Finland (Karling, 1933), and the tolerance to
seawater immersion shown by the terrestrial nemerteans *Geonemertes
agricola* (Crozier, 1917) and *G. nightingaleensis* (Brinkmann, 1947).
Coe (1904) held the view that the Bermudan species (*agricola*) was
clearly derived from marine rather than limnetic ancestors since the
island does not possess standing bodies of fresh water. That other
terrestrial forms (e.g., *G. dendyi*) are tolerant of prolonged freshwater submersion (Pantin, 1969) does not invalidate this reasoning,
since it can be reasonably explained by considering them as further
advanced towards a fully terrestrial habitat than their semi-littoral
relatives.

Moore and Gibson (unpublished) extend their theory of freshwater prosorhochmid ancestry by suggesting that, since the different
representatives of the terrestrial genus *Geonemertes* occur from
widely spaced and isolated localities, each might give rise to a distinct freshwater form providing suitable environmental conditions
prevail.

The routes whereby the heteronemertean and bdellonemertean

species have gained access to freshwater environments cannot be similarly demonstrated, but the absence of terrestrial examples from these orders may be indicative of an estuarine route.

No detailed ecological investigations have been carried out with freshwater nemerteans, but Child (1901) and Poluhowich (1968) have recorded some details for the North American species, *Prostoma rubrum*. Both authors found that the worms were particularly common amongst filamentous algae but also occurred regularly upon other aquatic plants. Poluhowich found that his Connecticut population seemed to be restricted to the pond-marsh interface where there was an abundance of decaying organic material. This area varies in temperature through the year from 2°C to 19°C, but retains a more or less constant acidity of pH 5. Even when the remainder of the pond is iced over during the winter months, *Prostoma* can still be obtained in plentiful numbers from the pond margins, and there is no evidence of the seasonal migration to deeper waters reported by Child for a Chicago population.

Stiasny-Wijnhoff (1938), in her review of the genus *Prostoma*, lists fully the global distribution of the various species found prior to that date. Among the later articles dealing with the occurrence of *Prostoma* species from various geographical localities may be cited those of Rioja (1941), Cordero (1943) and Corrêa (1951) from the Central and Southern Americas, Berg (1948), Gontcharoff and Deroux (1949), Lepori (1949) and Gibson and Young (1971) from Europe, and Gibson and Moore (1971) from Tasmania.

Reports dealing with other freshwater genera include du Bois-Reymond Marcus (1948) on *Siolineus* from Brazil, Beauchamp (1928) on *Planolineus* from Europe, Moore and Gibson (1972, 1973) on *Campbellonemertes* and *Potamonemertes* from Campbell Island and New Zealand respectively, and Blanchard (1847) on *Malacobdella* from Chile.

Terrestrial nemerteans

All the terrestrial species currently recognised belong to the single hoplonemertean genus *Geonemertes*, although Pantin (1969) points out that such a range of morphological differences occurs between the various forms that the group may not be a 'natural' one. Friedrich (1955), for these very reasons, subdivided the terrestrial nemerteans into three genera (*Geonemertes, Leptonemertes, Neonemertes*). Pantin (1969), however, does not accept this division and considers that there is insufficient evidence at present to determine accurately where true relationships exist between the several species. He does, nevertheless, point out that within the genus as a whole some of the species that are obviously closely related can be placed in one of two major groups, the *pelaensis* and Australian forms.

There seems to be little doubt that the terrestrial nemerteans have evolved from marine rather than freshwater ancestors. Pantin (1969) believed that this evolutionary sequence had occurred independently in the various widely separated localities, and was in some way related to the changes that took place in sea level and climate during the Pleistocene period.

Terrestrial nemerteans live in conditions of dampness, high humidity and reduced light intensity. They are found amongst leaf litter, beneath stones or fallen logs, or under the bark of decaying trees. One example, *Geonemertes arboricola* (Punnett, 1907) from the Seychelle Islands, lives in the moist humus between the leaf bases of the pine *Pandanus*.

The worms appear to possess far more precise environmental requirements than do terrestrial flatworms (Dr J. Moore, personal communication), and in the British Isles *Geonemertes dendyi*, an introduced Australian species, has never been found beyond a few miles from the coast, and mostly to the warm side of the mean 5°C January isotherm (Pantin, 1961b).

The genus as a whole is found from below low-tide level (*Geonemertes agricola*) up to heights of 4000 to 5000 feet above sea level (*G. australiensis*, *G. hillii*), as well as a considerable distance inland (Crozier, 1917; Hickman, 1963; Pantin, 1969). In their general morphology they are very similar to the marine prosorhochmids, but certain characters, such as the rhynchocoelic vascular plugs and extensive nephridial system, might be structurally modified in relation to the adoption of terrestrial habits. A parallel can be drawn between these forms and the true brackish-water species discussed earlier.

The geographical distribution of the various species is given in Table 6.

Bathypelagic nemerteans

All the bathypelagic nemerteans belong to the polystyliferous hoplonemertean tribe Pelagica, and possess broad flattened bodies quite unlike those of the more conventional littoral species (see Chapter 1). The polystyliferous Reptantia (e.g., *Drepanophorus*, *Uniporus*), whose general organisation, with the exception of the proboscis apparatus, is closely similar to that of the monostyliferous hoplonemerteans, are believed to represent the ancestral and more advanced forms from which the pelagic species have evolved (Brinkmann, 1917; Coe, 1926).

Pelagic nemerteans are entirely oceanic, various species having

Table 6
The geographical distribution of terrestrial nemerteans (from Pantin,
1961b, 1969).

Geonemertes agricola	Bermuda
Geonemertes arboricola	Seychelle Islands
Geonemertes australiensis	S.E. Australia, Tasmania
Geonemertes chalicophora	European greenhouses, Azores?, Madeira?
Geonemertes dendyi	S.W. Australia, European greenhouses,* British Isles*
Geonemertes hillii	New South Wales
Geonemertes nightingaleensis	Nightingale Island (Tristan da Cunha)
Geonemertes novaezealandiae	New Zealand
Geonemertes pantini	New Zealand
Geonemertes pelaensis	Pelew and Caroline Islands, Samoa, Celebes, Kei, Upolu, Ceylon
Geonemertes rodericana	Rodrigues Island (Indian Ocean)

* Introduced from Australia

Other 'species' that are either not or incompletely described are omitted from the
list above. These include *caeca, graffi, micholitzi, peradeniya, spirospermia* and
vinsoni.

been recorded from all the major water masses of the world, and
none have been found in shallower seas or in sheltered conditions.
More than 50 per cent of the described species are known from only
single localities, and few are known with wide geographic ranges.
The most well-distributed forms are *Dinonemertes investigatoris*,
from the North Atlantic to the Indian Ocean, *Nectonemertes mirabilis* (throughout the North Atlantic), *N. pelagica* (from 52°N to 12°S
in the Pacific), *N. minima* and *N. primitiva* (from 57°N to 35°S in the
Atlantic), and *Pelagonemertes rollestoni* (from the equatorial Atlantic, off Africa, to 4°N in the Indian Ocean and 50°S in the Atlanto-
Pacific regions) (Coe, 1926).

Most species occur only at extreme depths. Coe (1936) lists only
one specimen out of seventy-nine trawled from less than 1000 metres
off Bermuda, and recoveries from deeper than 2000 metres were
made during the Michael Sars Expedition of 1910. At oceanic
depths below about 1200 metres environmental conditions, particularly oxygen and phosphate content, temperature and salinity,
become stabilised, and there is every reason to suppose that the deep-
water species potentially possess wide distributional limits both
vertically and horizontally. Geographic dispersion, however, is
entirely dependent upon the deep oceanic currents, for none of the
bathypelagic forms are strong or active swimmers. Coe (1945a)
comments that it would take forty to fifty years for South Atlantic
species to be carried northwards into the vicinity of Bermuda.

Brinkmann (1917), in his monograph on the pelagic nemerteans, showed that *Nectonemertes mirabilis* was mostly confined to water layers where the temperature did not exceed 6°C and the salinity was 35‰ or less. In the North Atlantic these conditions occur at depths of about 1500 metres off the west coast of Ireland, up to quite near the surface in the neighbourhood of Newfoundland, with the result that in the western Atlantic the species has been found at much shallower depths (500 metres) than in the east. The lower limits of *Nectonemertes mirabilis* or, in fact, of any species, are not known, but most bathypelagic nemerteans have been recovered from depths of between 1000 and 2000 metres. A few forms are found in somewhat shallower conditions; *Balaenanemertes lobata* has been recovered from only 400 metres, and species of *Armaueria*, *Parabalaenanemertes*, *Probalaenanemertes* and *Pelagonemertes* have been trawled from depths of 600 to 800 metres. At the other extreme, eight species are so far known from depths of 3000 metres or more, with *Chuniella elongata* occurring in hauls made from 4000 metres.

Bathypelagic nemerteans are not common animals. Calculations based on figures given by Coe (1945a) relating to the Bermuda Oceanographic Expeditions of Dr William Beebe, made during the years 1929–31, suggest that a mean density of one specimen per 50,000 cubic metres of sea water trawled would be an optimistic figure. One animal was found, on average, for every eight hours of trawling.

It is hardly surprising that our knowledge about the bathypelagic hoplonemerteans, apart from their anatomy, is minimal, and Coe's (1926) statement that '. . . The reader can hardly fail to . . . be impressed with the opportunities awaiting future investigations into the pelagic life of those vast areas of the oceans as yet unexplored or but superficially explored . . .' is no less applicable today.

The known geographical distributions of the various genera is shown in Table 7.

Commensal and parasitic nemerteans

Most of the reports relating to commensal or parasitic habits in nemerteans have been of monostyliferous hoplonemertean species (*Carcinonemertes*, *Coenemertes*, *Emplectonema*, *Gononemertes*, *Nemertopsis*, *Tetrastemma*), the only exceptions being the monotypic heteronemertean genus *Uchidana*, and the monogeneric bdellonemertean order, with four species.

In most instances a definite relationship between nemertean and host has not been established, this being particularly true for the various *Tetrastemma* forms. Bürger (1897–1907) lists five species,

Table 7
The geographical distribution of bathypelagic nemertean genera
(from Coe, 1926, 1945a).

Armaueria NA	*Neuronemertes* EP
Balaenanemertes I, NA	*Pachynemertes* NA
Bürgeriella NA	*Parabalaenanemertes* NA
Calonemertes SA	*Paradinonemertes* NA
Chuniella EA, I, NA	*Pelagonemertes* EA, EP,
Crassonemertes NA	I, NA, NP, SA, SP
Cuneonemertes EP	*Pendonemertes* NA
Dinonemertes EP, I, NA	*Phallonemertes* NA
Gelanemertes NA	*Planktonemertes* EP, SA
Mergonemertes I	*Planonemertes* EP, NA
Mononemertes NA	*Plionemertes* EP
Nannonemertes I	*Plotonemertes* NA
Natonemertes NA	*Proarmaueria* NP
Nectonemertes EA, EP, NA,	*Probalaenanemertes* NA, SA
NP, SA	*Protopelagonemertes* NA, SA

EA=Equatorial Atlantic   EP=Equatorial Pacific   I=Indian Ocean   NA=
North Atlantic   NP=North Pacific   SA=South Atlantic   SP=South Pacific

under the generic name *Prostoma*, as parasites of tunicates, molluscs
or crustaceans; Marion (1874) had earlier recorded *Tetrastemma
kefersteinii* from the pharyngeal cavity of Mediterranean tunicates
and Coe (1905a) listed *T. caecum* as a possible parasite, also of tunicates. Other, free-living, *Tetrastemma* species are known to regularly
occur amongst colonies of tunicates, molluscs and other sessile
invertebrates (Coe, 1905a), and it seems probable that in most cases
reports have been merely of nemerteans occurring in those 'hosts'
purely by chance. Only Coe's and Marion's species show any morphological adaptations that might be related to commensal habits in
their loss of eyes and assumption of hermaphroditism. A definite
nemertean-host relationship, nevertheless, cannot be determined
merely on this basis, and it seems inadvisable at this juncture to draw
too firm a conclusion.

The same reasoning can be applied to Bürger's (1904b) report of
Nemertopsis actinophila occurring as a commensal beneath the pedal
discs of the anemones *Tealia davisii* and *Stomphia polaris*. Other
Nemertopsis species are completely errant in their habits.

Coenemertes caravela (Corrêa, 1966) and *Emplectonema kandai*
(Kato, 1939) have been found only on the decapods *Callianassa* and
Chelyosoma respectively, either crawling amongst the thoracic limbs
near the branchial chambers, or coiled up on various parts of the
body, but there is no evidence as to whether or not this association
is a permanent one.

Gononemertes parasita occurs in the atrium of tunicates, often with its head protruding into the pharyngeal cavity. This species has no errant relatives, and does show such morphological modifications as a reduction of the proboscis apparatus and multiplicity of the gonads that could be related to commensal habits (Bergendal, 1900; Brinkmann, 1927). With its head inserted into the pharyngeal water currents *Gononemertes* is well placed for obtaining a constantly replenished supply of food and oxygen. So far as is known, no adverse effects are felt by the host and the relationship between the two is one of commensalism.

The genus *Carcinonemertes*, with four recognised species, exhibits a very much more definite relationship with its various crab hosts; Humes (1942) regards *Carcinonemertes* as an ectohabitant, but there are grounds for arguing that the nemerteans are true parasites, particularly since they appear to feed only upon the eggs of the hosts. The four reported species, one of which is divided into two varieties, are:

Carcinonemertes carcinophila var. *carcinophila*: On galatheid, portunid and xanthid crabs; eight host species recorded; from Belgium, British Isles, France, Italy and the Atlantic United States.

Carcinonemertes carcinophila var. *imminuta*: Common on the portunid crab *Callinectes sapidus*, less frequently on leucosiid, calappid and xanthid crabs; fourteen host species recorded; from Brazil, Panama, Porto Rico, West Indies and Atlantic United States.

Carcinonemertes epialti: On the crabs *Euphylax dovii* and *Pugettia producta*; two host species recorded; from Peru and the Pacific United States.

Carcinonemertes mitsukurii: Common on four portunid crab species, also known from one grapsid species; from the Pacific (San Andreas, Hawaiian, Kingsmills and Society Islands), Japan, Singapore and Hong Kong.

Carcinonemertes coei: Known only from one female specimen of the portunid crab *Charybdis natator*; from Zanzibar.

The nemerteans are hardly ever found other than on female crabs; Humes (1942) records incidences of up to 63 per cent on gravid females compared with a maximum of only 1·15 per cent on ripe males and relates this to the fact that whereas immature crabs of both sexes and adult males moult at regular intervals, mature females do not. Any infection can thus be eliminated at each ecdysis.

The number of *Carcinonemertes* occurring on individual hosts is extremely variable, but can be very large. Humes (1942) recorded at least 1000 worms between the lamellae of both gill chambers of one crab, whereas Pearse (1949) found an average of only eighty-three *C. carcinophila* on specimens of *Callinectes sapidus*, and Coe (1902b) gave a slightly lower value of forty to sixty worms per host.

Immature *Carcinonemertes* are generally found on the gills of the host, where they often form capsules about themselves by secreting mucus to cement adjacent lamellae together. They remain within the gill chamber until the host becomes ovigerous, then migrating to the egg mass to deposit their own ova. *Carcinonemertes* attain sexual maturity only upon the host egg mass, and are never ripe whilst in the branchial region. Larval *Carcinonemertes*, on hatching, either remain amongst the host eggs and feed upon them, or leave the host and swim towards the surface waters through a positive phototactic behaviour. The worms that remain on the original host return to the gill chamber when the crab eggs hatch (Coe, 1902b; Humes, 1942; Davis, 1965).

The entire life cycle of *Carcinonemertes* is thus intimately linked with that of its host, although the nemerteans do not show a high degree of host specificity. Crab eggs and developing embryos are certainly utilised as food, but there is some doubt as to whether the nemerteans feed upon the host gills whilst they are immature. Coe (1902b) found severe gill damage which he believed to be due to *Carcinonemertes*, but other authors have not found similar evidence. The manner in which *Carcinonemertes* encapsulates itself on the gills is very reminiscent of the way in which many free-living species survive adverse conditions, and it may be that when in the gills, the nemerteans enter a non-feeding resting stage, awaiting the next ripe period of the host.

Far more portunid crab species have been recorded as hosts for *Carcinonemertes* than from any other family, and Humes (1942) explains this through the general habits of the Portunidae. The crabs are common in large numbers in shallower waters, where they swim actively near the surface; they are also entirely aquatic and do not live intertidally. In these circumstances the chances of infection by *Carcinonemertes* larvae must be high, the nemerteans not being exposed to problems such as desiccation and absence of other hosts. Although *Carcinonemertes* is not host specific, its whole life cycle does seem much more adapted for portunid infestation than any other crab family.

The heteronemertean *Uchidana parasita* is known only from the mantle cavity and spaces between shell and mantle of the bivalve mollusc *Mactra sulcataria*. It feeds upon host gill tissues, and can therefore be regarded as parasitic rather than commensal. No figures are available on infection incidence, but Iwata (1967) observes that the nemerteans are found only in the larger hosts.

The bdellonemertean group, with four species (*Malacobdella auriculae, M. grossa, M. japonica, M. minuta*) is entirely commensal within the mantle cavity of various bivalve molluscs. Apart from *Malacobdella grossa*, which has so far been reported from twenty-two different mollusc species (Gibson, 1968), the nemerteans are known only from single host species and localities. *Malacobdella grossa* occurs on both sides of the North Atlantic, and is also reported from the north-east Pacific, the Mediterranean and the North Sea.

Malacobdella does not in any way appear to affect its hosts, and does not feed upon host tissues (Gibson and Jennings, 1969). Its relationship with bivalves seems to be entirely one of commensalism, the molluscs affording shelter to the nemerteans and, by means of their ciliary water currents, providing an abundance of food and oxygen.

Gibson (1967) reported that a definite, although not absolute, correlation exists between both incidence of infection and number of multiple infections with the size of the host. Larger hosts tend to shelter larger *Malacobdella* and more multiple infections, and the overall infection incidence also increases with host size. In well-established host populations 60 per cent or more of the molluscs may harbour nemerteans. The maximum number of worms recovered from a single host is five, and multiple infections are nearly always composed entirely of small and young nemerteans. Most cases of infection (90 per cent or more) involve only a single *Malacobdella*.

Within its host *Malacobdella* can be found in any position in which it is exposed to inhalant water currents. Most commonly it occurs on the mantle wall, or on or between the gills, but it may be recovered from the ventral wall of the visceral mass or from within the host inhalant siphon.

The bdellonemerteans are the only nemertean group to show the development of a posterior sucker for attachment, and the formation of this structure poses interesting evolutionary problems. In other ways too, *Malacobdella* shows morphological modifications related to its mode of life, particularly in the loss of eyes, simplified nervous system and increased number of gonads. In its aberrant anatomy it is incapable of surviving for too long away from its host, being apparently unable to acquire food for itself. *Carcinonemertes*, although not so anatomically adapted (it has a reduction in eye number and much reduced proboscis) is certainly behaviourally adapted and dependent upon its crab hosts, and these two genera are perhaps the only nemertean types truly, on the one hand commensal, on the other parasitic, no other forms being so modified.

7

PHYLOGENETIC RELATIONSHIPS

Reference has been made in the preceding chapters to several points of similarity between nemerteans and turbellarian flatworms, and there can be little doubt, as Hyman (1951) states, that a close relationship exists between them. The general structure of both groups consists essentially of a block of parenchymatous tissue, enclosing the various body organs, externally bounded by concentric layers of musculature and epidermis. Both are also unsegmented and bilaterally symmetrical.

It is at the histological level that even more striking similarities are encountered. The ciliated epidermis of nemerteans, especially in those forms containing rhabdite-like cells, is very like that found in triclads and polyclads. Eyes, which occur in all turbellarian orders, bear a striking resemblance to those of nemerteans. The cephalic or frontal glands, so common in nemerteans, are additionally found in most acoel and several alloeocoel turbellarians, less frequently in the rhabdocoels, but not at all in polyclads and triclads.

In nemerteans the cephalic slits, grooves and pits form important sensory structures of the anterior body region, and homologous structures are present in many turbellarians. The cerebral organ of nemerteans, consisting of a ciliated cerebral canal terminating in a neuroglandular complex, can be fairly easily derived by the invagination of the triclad auricular grooves coming into contact with nervous and gland-cell components. Statocysts, known within nemerteans only from the single genus *Ototyphlonemertes*, are characteristic of the more primitive Turbellaria, occurring in most acoels, many alloeocoels and the rhabdocoel genus *Catenula*. Sensory epidermal spines or cilia, recorded from such hoplonemertean genera as *Ototyphlonemertes* and *Prostoma*, are also common in many rhabdocoels.

188 Nemerteans

A close resemblance is found between the basic plan of the nemertean central nervous system and that occurring in turbellarians, and both groups possess a protonephridial excretory system comprised of distinctive ducts and flame cells.

The pseudometameric arrangement of the gonads and intestinal diverticula is repeated in several of the maricolan triclads, notably in *Procerodes lobata*.

The proboscis of nemerteans, although differing quite extensively from any turbellarian structure, does possess some similarities in common with the proboscis of kalyptorhynchian rhabdocoels. Such a link, first postulated by Salensky (1884), was not generally accepted until after Wijnhoff's (1914) paper on the nemertean proboscidial system, and no attempt has yet been made from a phylogenetic aspect to differentiate between the eukalyptorhynchian undivided proboscis and the bifid structures of the Schizorhynchia.

At the embryological level similarities are seen in the spiral type of cleavage possessed by both groups. The lower turbellarian orders (the archoophoran grade), however, which includes acoels, macrostomid rhabdocoels and certain polyclads, show a spiral cleavage pattern that is more primitive than that of nemerteans (Beklemishev, 1963). Their cleavage shows an homoquadrant orientation, whereas in nemerteans blastomere differentiation is of common occurrence due to the unequal division of micro- and macromeres.

Physiological evidence can also be provided that indicates a close relationship between the two groups. The serological studies of Schepotieff (1912) showed that whereas *Cerebratulus* antiserum reacted positively with a polyclad extract, it did not with that of annelids. This suggests that the affinities of nemerteans are close to turbellarians, whilst no comparable relationship with annelids can be demonstrated.

The digestive physiology of the two groups is very similar, both relying upon a combination of intra- and extracellular processes that utilise the same enzyme groups at parallel stages.

Several nemertean species show remarkable powers of regeneration; so too do many turbellarians, particularly examples of freshwater triclads and, to a lesser extent, the rhabdocoels.

There is thus an overwhelming volume of evidence to link nemerteans with flatworms. However, the presence in nemerteans of a blood vascular system and an anus, and the greater degree of elaboration shown by many structures with turbellarian homologues, leads to the inevitable conclusion that the nemertean level of acoelomate organisation is considerably more advanced than that of flatworms. An important pointer to an early dichotomy of the two groups is provided by a comparison of their reproductive apparatus. Turbellarians are predominantly hermaphroditic with complex genital arrangements, whereas nemerteans are mostly bisexual and only extremely rarely show any tendency to genital elaboration. This situation is much more readily explained by an early separation in the evolutionary lineage than by a later degeneration of more complicated structures. If this inference is taken in conjunction with other morphological and embryological evidence, it can be suggested that nemerteans have most probably evolved from an early rhabdocoel type of turbellarian, their line splitting off from platyhelminths soon after the development of a rudimentary proboscis-like structure. This suggestion is supported by the living representatives of the rhabdocoels with a proboscis, the kalyptorhynchians.

An even closer link between rhabdocoels and nemerteans is provided by the macrostomid turbellarian *Haplopharynx rostratus*. This species, in its general morphology, resembles nemerteans more closely than any other known platyhelminth, even to the point of possessing a supraterminal anal pore (Karling, 1965).

Hyman (1951) regards nemerteans as representing the culmination in development of this type of construction, and certainly no more advanced acoelomate organisms are known. This does not necessarily mean, however, that nemerteans represent the peak of an evolutionary cul-de-sac. Recent workers (Jensen, 1960, 1963; Willmer, 1970) have proposed that deuterostomes can be derived from nemerteans, and that the group is therefore on the direct line of evolution leading to man himself. Jensen (1963) states that a satisfactory ancestral line to the myxinoids does not exist prior to the Ordovician period, and that his hoplonemertean hypothesis, somewhat different to those postulated earlier by Hubrecht (1883, 1887b) and Macfarlane (1918), provides a more logical link with primitive life than does the currently more widely accepted protochordate theory. Jensen's hypothesis can best be illustrated by listing his evolutionary derivations from hoplonemertean structures, as given in his 1960 paper.

Many of the analogies drawn appear reasonable and would not require severe modification from the nemertean condition; such structures include the anus, vascular system, solid nerves and eyes. However, equally strong arguments could be put forward in favour of myxinoid derivation from similar structures in other invertebrate groups. Using many of Jensen's arguments one could even derive the ancestral myxinoids direct from turbellarians without involving nemerteans at all.

Other links in the theory are at best difficult to accept. For example, close analogies are drawn between the stylet basis and

Hoplonemerteans	Myxinoids
Secondary mouth	Nasopharyngeal duct, mouth, gill slits
Glandular proboscis elements	Glandular hypophysis of nasopharyngeal duct
Muscular proboscis elements	Anterior muscular somites
Proboscis sheath	Notochord sheath
Rhynchocoel fluid	Notochord contents
Proboscis retractor muscle	Notochord strand
Stylet basis	Skeletal cartilage
Stylet	Horny teeth
Dorsal blood vessel	Dorsal aorta
Closed vascular system	Closed vascular system
Intestinal caecum	Liver
Anus	Anus
Lateral eyes, direct retina eyes	Lateral eyes, pineal eye
Cerebral organs	Labyrinth organs
Subepidermal organs	Neuromast cells
Frontal organ	Olfactory organs, dorsal neural tube
Solid nerves	Solid nerves
Dorsal ganglia	Dorsal sensory alar plates
Ventral ganglia	Ventral motor basal plates

myxinoid skeletal material, the stylet and horny pharyngeal teeth, rhynchocoel fluid and notochord contents, and other proboscis components with different parts of the notochord. Certainly so far as the stylet apparatus is concerned we have absolutely no evidence to show whether or not any similarity exists between it and skeletal material, nor does Jensen's reference to the muscular proboscis sheath of pelagic hoplonemerteans acting as a 'stiffening aid' in locomotion at all account either for the sluggish swimming shown by these nemerteans or the need of such a muscular arrangement in animals whose parenchyma has been modified for reasons of buoyancy. Similarly, he fails to indicate how such basic requirements as gill slit vascular supplies, coelomic cavities, or somitic replication of the main body musculature could have evolved. Such statements as that denoting the ventral rhynchocoel epithelium as the 'pumping agent' for blood circulation are based on no evidence whatsoever, and can only be regarded as guesswork suitable to the overall hypothesis.

A dependence upon selected specific points, rather than the overall picture, forms the basis for Willmer's (1970) theories on postnemertean evolution, although he does not specify hoplonemerteans so much as rely principally on the heteronemertean *Lineus ruber* to provide homologous structures. Willmer, in fact, outlines a far more tortuous route to the deuterostome level of organisation than does Jensen. In this respect Willmer would perhaps have been more convincing if only he had not so obviously selected points in favour of his suggestions (using *Lineus* where possible, but otherwise drawing upon any other nemertean species that happens to fit), and largely ignored many more that are unsuitable. Unfortunately, in the arguments put forward, such errors as regarding the foregut of *Lineus* as 'segmented', and stressing the importance of the ventral foregut folding as a precursor of the endostyle, leads the reader to view other interpretations with caution. The foregut of *Lineus* is undoubtedly folded, as it is in many other species, but there is no permanent ventral groove as Willmer infers. In fact in an animal requiring expansile properties in order to ingest large morsels of food, which nemerteans frequently do, folding of the gut wall is a prerequisite. With the proboscis apparatus in a dorsal position, expansion is somewhat limited to this side of the gut with the consequence that any dilation that does occur must do so laterally or ventrally. The tendency for epithelial folding to be reduced or absent from the dorsal regions of the foregut can thus be easily and logically explained without the need for other suggestion.

Such aspects of nemertean biology as their spiral cleavage and other embryological details Willmer hardly mentions, nor does he seem to appreciate that in many respects rhabdocoels would be as applicable to his arguments as are nemerteans. Willmer clearly wrote his book to stimulate thought on the evolutionary ancestry of vertebrates and many of his other suggestions are more acceptable. In fact his overall hypothesis cannot be entirely disproved; perhaps what is more important is that it spotlights nemerteans from a new and potentially interesting angle.

APPENDIX

THE CLASSIFICATION OF NEMERTEANS

The morphological relationships of nemerteans are so far from being well understood that in many instances taxonomic grouping cannot be achieved with any degree of certainty. Friedrich (1955, 1960), amongst others, has carefully avoided the use of family names in his reviews of the monostyliferous hoplonemerteans and heteronemerteans. Whilst agreeing with Friedrich's caution, it is nevertheless useful to have at least some idea of the systematic relationships of the various nemertean genera. Accordingly familial names are used in this classification; the reader should, however, be aware of the limitations shown by this arrangement. Genera of particularly uncertain affinities are listed in brackets.

PHYLUM: Nemertea, Nemertini, Nemertinea or Rhynchocoela

Class I: ANOPLA

Order 1: PALAEONEMERTEA (PALAEONEMERTINI)
1. Family: CARINOMIDAE. *Carinoma*, (*Carinomella*)
2. Family: CEPHALOTHRICIDAE. *Cephalothrix, Cephalotrichella, Procephalothrix*
3. Family: HUBRECHTIDAE. *Hubrechtella, Hubrechtia*
4. Family: TUBULANIDAE. *Callinera, Carinesta, Carinina, Tubulanus*

Order 2: HETERONEMERTEA (HETERONEMERTINI)[1]

[1] The five heteronemertean families listed are equivalent to the groups recognised by Friedrich (1960), and are not necessarily natural divisions. The family names 'Poliopsiidae' and 'Pussylineidae' are unofficial, but have been coined in accordance with the International Rules of Taxonomy.

G

1. Family: BASEODISCIDAE. *Baseodiscus, Parapolia*
2. Family: LINEIDAE. *Cerebratulides, Cerebratulus, Chilineus, Corsoua, Diplopleura, Dushia, Euborlasia, Evelineus, Flaminga, Gorgonorhynchus, Heterolineus, Hinumanemertes, Lineopsis, Lineus, Micrella, Micrura, Micrurina, Parborlasia, Pontolineus, Pseudolineus, Siolineus, Uchidana*
3. Family: 'POLIOPSIIDAE'. *Poliopsis*
4. Family: 'PUSSYLINEIDAE'. *Lineopsella, Pussylineus*
5. Family: VALENCINIDAE. *Cephalomastax, Oxypolella, Oxypolia, Paralineus, Planolineus, Valencinia, Valencinura, Zygeupolia*

Class II: ENOPLA

Order 3: HOPLONEMERTEA (HOPLONEMERTINI)
Suborder 1: MONOSTYLIFERA

1. Family: AMPHIPORIDAE. *Africanemertes, (Alaonemertes), (Algonemertes), Amphiporella, Amphiporus, (Austroprostoma), Communoporus, Correanemertes, Cratenemertes, Duosnemertes, (Gononemertes), Gurjanovella, Intestinonemertes, Itanemertes, Korotkevitschia, Nipponemertes, (Obürgeria), Paramphiporus, Poikilonemertes, Poseidonemertes, Proneurotes, Sagaminemertes, Tagonemertes, Valdivianemertes, Zygonemertes*
2. Family: CARCINONEMERTIDAE. *Carcinonemertes*
3. Family: EMPLECTONEMATIDAE. *Emplectonema, Nemertes, Nemertopsella, Nemertopsis, Paranemertes*
4. Family: OTOTYPHLONEMERTIDAE. *Otonemertes, Ototyphlonemertes*
5. Family: PROSORHOCHMIDAE. *Acteonemertes, (Arenonemertes), Campbellonemertes, (Dichonemertes), (Friedrichia), Geonemertes, Oerstedia, Oerstediella, Paroerstedia, Potamonemertes, Prosorhochmus, (Sacconemertella), (Sacconemertes), (Sacconemertopsis)*
6. Family: TERTASTEMMATIDAE. *Antarctonemertes, (Atyponemertes), Nemertellina, Nemertellopsis, (Prosadenoporus), Prostoma, Prostomatella, Prostomiopsis, Tetrastemma*
Unplaced: *Amphinemertes, Arctonemertes, Coenemertes, Dananemertes*

Suborder 2: POLYSTYLIFERA
Tribe 1: REPTANTIA
   1. Family: BRINKMANNIDAE. *Brinkmannia*
   2. Family: DREPANOPHORIDAE. *(Drepanogigas), (Dre-panophorella), Drepanophorina, Drepanophorus, (Para-drepanophorus), Punnettia, (Uniporus), (Wijnhoffella)*
   3. Family: SIBOGANEMERTIDAE. *Siboganemertes*
Tribe 2: PELAGICA
   1. Family: ARMAUERIIDAE. *Armaueria, (Cuneonemertes), Proarmaueria*
   2. Family: BALAENANEMERTIDAE. *Balaenanemertes*
   3. Family: BÜRGERIELLIDAE. *Bürgeriella*
   4. Family: CHUNIELLIDAE. *Chuniella*
   5. Family: DINONEMERTIDAE. *Dinonemertes, Paradinone-mertes, Planonemertes, Plionemertes*
   6. Family: NECTONEMERTIDAE. *Nectonemertes*
   7. Family: PELAGONEMERTIDAE. *(Gelanemertes), Nanno-nemertes, Natonemertes, Parabalaenanemertes, Pela-gonemertes, Probalaenanemertes*
   8. Family: PHALLONEMERTIDAE. *Phallonemertes*
   9. Family: PLANKTONEMERTIDAE. *Crassonemertes, Mer-gonemertes, Mononemertes, Neuronemertes, Pachy-nemertes, Planktonemertes*
  10. Family: PROTOPELAGONEMERTIDAE. *Calonemertes, Pen-donemertes, Plotonemertes, Protopelagonemertes*

Order 4: BDELLONEMERTEA (BDELLONEMERTINI, BDELLOMORPHA)
   1. Family: MALACOBDELLIDAE. *Malacobdella*

BIBLIOGRAPHY

Articles marked with an asterisk have not been seen in the original

ALVARADO, R., 1956. Istologia delle formazioni epidermiche e dermomuscolari del *Cerebratulus marginatus* Ren. *Pubbl. Staz. zool. Napoli*, **28**, 1–11.

ARNOLD, G., 1898. Zur Entwicklungsgeschichte des *Lineus gesserensis* O. F. Müller (*L. obscurus* Barrois). *Trudy imp. S-peterb. Obshch. Estest.*, **28**, 21–30.

BACQ, Z. M., 1936. Les poisons des Némertiens. *Bull. Acad. r. Belg. Cl. Sci.*, Ser. 5, **22**, 1072–9.

BACQ, Z. M., 1937. L' 'amphiporine' et la 'némertine' poisons des vers némertiens. *Archs int. Physiol.*, **44**, 190–204.

BACQ, Z. M., 1947. L'acétylcholine et l'adrénaline chez les invertébrés. *Biol. Rev.*, **22**, 73–91.

BAIRD, W., 1866. Description of a new species of Monoecious worm, belonging to the class Turbellaria and genus *Serpentaria*. *Proc. zool. Soc. Lond.*, 101–2.

BALFOUR, W. E., and WILLMER, E. N., 1967. Iodine accumulation in a nemertine, *Lineus ruber*. *J. exp. Biol.*, **46**, 551–6.

BARROIS, J., 1877. Mémoire sur l'embryologie des Némertes. *Annls Sci. nat.*, Ser. 6, Zool., **6**, no. 3, 1–232.

BASTER, J., 1762. *Opuscula subseciva*. Harlem, **1**, 1–154.

BEAUCHAMP, P. de, 1928. Une Hétéronémerte d'eau douce à Buitenzorg. *Bull. Soc. zool. Fr.*, **53**, 62–7.

* BEKLEMISHEV, K. V., 1955. Predatory nemertines. *Priroda, Mosk.*, **9**, 108–9.

BEKLEMISHEV, V. N., 1963. On the relationship of the Turbellaria to other groups of the animal kingdom. In: *The Lower Metazoa, Comparative Biology and Phylogeny*. Ed. E. C. Dougherty, Univ. Calif. Press, pp. 234–44.

BERG, K., 1948. Biological studies on the River Susaa. *Folia limnol. scand.*, no. 4, 318 pp.

BERGENDAL, D., 1900. Über ein Paar sehr eigenthümliche nordische Nemertinen. *Zool. Anz.*, **23**, 313–28.

BERGENDAL, D., 1902. Zur Kenntnis der nordischen Nemertinen. 2. Eine der construirten Urnemertine entsprechende Palaeonemertine aus dem Meere der schwedischen Westküste. *Zool. Anz.*, **25**, 421–32.

BERGENDAL, D., 1903. Über 'Sinnesgrübchen' im Epithel des Vorderkopfes bei *Carinoma armandi* sp. McInt. (Oudemans) nebst einigen systematischen Bemerkungen über die Arten dieser Gattung. *Zool. Anz.*, **26**, 608–19.

198   Bibliography

BIANCHI, S., 1969a. On the neurosecretory system of *Cerebratulus marginatus* (Heteronemertini). *Gen. comp. Endocr.*, **12**, 541–8.

BIANCHI, S., 1969b. The histochemistry of the neurosecretory system in *Cerebratulus marginatus* (Heteronemertini). *Gen. comp. Endocr.*, **13**, 206–10.

BIERNE, J., 1962a. L'autodifférenciation du bourgeon rhynchogène de régénération chez *Lineus ruber* Müller. *C. r. hebd. Séanc. Acad. Sci., Paris*, **256**, 2833–5.

BIERNE, J., 1962b. La régénération de la trompe chez les Némertes. *Bull. biol. Fr. Belg.*, **96**, 481–504.

BIERNE, J., 1964. Maturation sexuelle anticipée par décapitation de la femelle chez l'Hétéronémerte *Lineus ruber* Müller. *C. r. hebd. Séanc. Acad. Sci., Paris*, **259**, 4841–3.

BIERNE, J., 1966. Localisation dans les ganglions cérébroïdes du centre régulateur de la maturation sexuelle chez la femelle de *Lineus ruber* Müller (Hétéronémertes). *C. r. hebd. Séanc. Acad. Sci., Paris*, **262**, 1572–5.

BIERNE, J., 1967a. Viabilité, stabilité phénotypique, croissance et régénération postérieure de chimères interspécifiques obtenues par la greffe chez des Nemertiens Lineidae adultes. *C. r. hebd. Séanc. Acad. Sci., Paris*, **264**, 1080–3.

BIERNE, J., 1967b. Sur le contrôle endocrinien de la différenciation du sexe chez la Némerte *Lineus ruber* Müller. La masculinisation des ovaires des chimères hétérosexuées. *C. r. hebd. Séanc. Acad. Sci., Paris*, **265**, 447–50.

BIERNE, J., 1968. Facteur androgène et différenciation du sexe chez la Némerte *Lineus ruber* Müller. L'effet 'free-martin' dans la parabiose hétérosexuée. *C. r. hebd. Séanc. Acad. Sci., Paris*, **267**, 1646–8.

BIERNE, J., 1970a. Recherches sur la différenciation sexuelle au cours de l'ontogenèse et de la régénération chez le némertien *Lineus ruber* (Müller). *Annls Sci. nat., Zool.*, **12**, 181–298.

BIERNE, J., 1970b. Influence des facteurs hormonaux gonado-inhibiteur et androgène sur la différenciation sexuelle des parabiontes hétérosexués chez un némertien. *Annls Biol.*, **9**, 395–400.

BIERNE, J., 1970c. Aspects expérimentaux de la différenciation sexuelle chez *Lineus ruber* (Hétéronémertes). *Bull. Soc. zool. Fr.*, **95**, 529–43.

BLAINVILLE, H. M. de, 1828. Vers et Zoophytes. In: *Dictionnaire des Sciences Naturelles*, Ed., F. G. Levrault, **57**, 573–7.

BLANCHARD, E., 1847. Sur l'organisation des vers. *Annls Sci. nat.*, Ser. 3, **8**, 119–49.

BLANCHARD, E., 1849. Recherches sur l'organisation des vers. *Annls Sci. nat.*, Ser. 3, **12**, 5–68.

BÖHMIG, L., 1898. Beiträge zur Anatomie und Histologie der Nemertinen (*Stichostemma graecense* (Böhmig), *Geonemertes chalicophora* (Graff)). *Z. wiss. Zool.*, **64**, 479–564.

BORLASE, W., 1758. *The Natural History of Cornwall.* Jackson, Oxford, 326 pp.

BOSC, L. A. G., 1802. Histoire naturelle des Vers. Paris, **1**, 256–62.

BRINKMANN, A., 1912. *Bathynectes murrayii* n. gen. n. sp. Eine neue bathypelagische Nemertine mit äusseren männlichen Genitalien. *Bergens Mus. Årb.*, no. 9, 1–9.

BRINKMANN, A., 1917. Die Pelagischen Nemertinen. *Bergens Mus. Skr.*, **3**, 1–194.

BRINKMANN, A., 1927. *Gononemertes parasita* und ihre Stellung im System. *Nyt Mag. Naturvid.*, **65**, 57–81.

BRINKMANN, A., Jr., 1947. A new land nemertean from the Tristan da Cunha group, and a survey of the genus *Geonemertes*. *Results Norw. scient. Exped. Tristan da Cunha*, no. 15, 1–24.

BRUNBERG, L., 1959. *Emplectonema bocki* n. sp., a hoplonemertean epizoic on *Funiculina quadrangularis* (Pallas). *Vidensk. Meddr. dansk naturh. Foren.*, 119, 59–66.

BRUNBERG, L., 1964. On the nemertean fauna of Danish waters. *Ophelia*, 1, 77–111.

BÜRGER, O., 1893. Südgeorgische und andere exotische Nemertinen. *Zool. Jb.*, 7, 207–40.

BÜRGER, O., 1895. Die Nemertinen des Golfes von Neapel und der Angrenzenden Meeres-Abschnitte. *Fauna Flora Golf. Neapel*, 22, 1–743.

BÜRGER, O., 1897–1907. Nemertini (Schnurwürmer). In: H. G. Bronn's *Klassen und Ordnungen des Tier-Reichs*, vol. 4, Suppl., 542 pp.

BÜRGER, O., 1904a. Nemertinen. *Résult. Voyage S.Y. Belgica, Zool.*, 1–10.

BÜRGER, O., 1904b. Die Nemertinen. *Fauna arct.*, 3, 55–64.

CANTELL, C.-E., 1966a. The devouring of the larval tissues during the metamorphosis of pilidium larvae (Nemertini). *Ark. Zool.*, 18, 489–92.

CANTELL, C.-E., 1966b. Some developmental stages of the peculiar nemertean larva *pilidium recurvatum* Fewkes from the Gullmarfjord (Sweden). *Ark. Zool.*, 19, 143–7.

CANTELL, C.-E., 1969. Morphology, development, and biology of the pilidium larvae (Nemertini) from the Swedish west coast. *Zool. Bidr. Upps.*, 38, 61–111.

CAULLERY, M., 1908. Sur une anomalie de la trompe chez un némertien (*Tetrastemma candidum* O.F.M.). *C. r. Séanc. Soc. Biol.*, 64, 738–40.

CHILD, C. M., 1901. The habits and natural history of *Stichostemma*. *Am. Nat.*, 35, 975–1006.

CLARK, R. B., 1964. *Dynamics in Metazoan Evolution. The Origin of the Coelom and Segments*. Clarendon, Oxford, 313 pp.

CLARK, R. B., and COWEY, J. B., 1958. Factors controlling the change of shape of certain nemertean and turbellarian worms. *J. exp. Biol.*, 35, 731–48.

COE, W. R., 1895. On the anatomy of a species of nemertean (*Cerebratulus lacteus* Verrill), with remarks on certain other species. *Trans. Conn. Acad. Arts Sci.*, 9, 479–514.

COE, W. R., 1899a. On the development of the pilidium of certain nemerteans. *Trans. Conn. Acad. Arts Sci.*, 10, 235–62.

COE, W. R., 1899b. Notes on the times of breeding of some common New England nemerteans. *Science, N.Y.*, 9, 167–9.

COE, W. R., 1899c. The maturation and fertilisation of the egg of *Cerebratulus*. *Zool. Jb.*, 12, 425–76.

COE, W. R., 1901. Papers from the Harriman Alaska Expedition. XX. The nemerteans. *Proc. Wash. Acad. Sci.*, 3, 1–110.

COE, W. R., 1902a. The genus *Carcinonemertes*. *Zool. Anz.*, 25, 409–14.

COE, W. R., 1902b. The nemertean parasites of crabs. *Am. Nat.*, 36, 431–50.

COE, W. R., 1902c. The nemerteans of Porto Rico. *Bull. U.S. Fish Commn.*, 20, 223–9.

COE, W. R., 1904. The anatomy and development of the terrestrial nemertean (*Geonemertes agricola*) of Bermuda. *Proc. Boston Soc. nat. Hist.*, 31, 531–70.

COE, W. R., 1905a. Nemerteans of the west and northwest coasts of America. *Bull. Mus. comp. Zool. Harv.*, 47, 1–318.

COE, W. R., 1905b. Synopses of North American invertebrates. XXI. The nemerteans. *Am. Nat.*, 39, 425–47.

COE, W. R., 1906. A peculiar type of nephridia in nemerteans. *Biol. Bull. mar. biol. Lab., Woods Hole*, **11**, 47–52.

COE, W. R., 1920. Sexual dimorphism in nemerteans. *Biol. Bull. mar. biol. Lab., Woods Hole*, **39**, 36–58.

COE, W. R., 1926. The pelagic nemerteans. *Mem. Mus. comp. Zool. Harv.*, **49**, 1–244.

COE, W. R., 1927a. Adaptations of the bathypelagic nemerteans. *Am. Nat.*, **61**, 345–52.

COE, W. R., 1927b. The nervous system of pelagic nemerteans. *Biol. Bull. mar. biol. Lab., Woods Hole*, **53**, 123–38.

COE, W. R., 1929a. The excretory organs of terrestrial nemerteans. *Biol. Bull. mar. biol. Lab., Woods Hole*, **56**, 306–11.

COE, W. R., 1929b. Regeneration in nemerteans. *J. exp. Zool.*, **54**, 411–59.

COE, W. R., 1930a. Unusual types of nephridia in nemerteans. *Biol. Bull. mar. biol. Lab., Woods Hole*, **58**, 203–16.

COE, W. R., 1930b. Two new species of nemerteans belonging to the family Cephalotrichidae. *Zool. Anz.*, **89**, 97–103.

COE, W. R., 1930c. The peculiar nephridia of the nemerteans of the genus *Cephalothrix*. *Zool. Anz.*, **89**, 103–8.

COE, W. R., 1930d. Asexual reproduction in nemerteans. *Physiol. Zoöl.*, **3**, 297–308.

COE, W. R., 1930e. Regeneration in nemerteans. II. Regeneration of small sections of the body split or partially split longitudinally. *J. exp. Zool.*, **57**, 109–44.

COE, W. R., 1931. A new species of nemertean (*Lineus vegetus*) with asexual reproduction. *Zool. Anz.*, **94**, 54–60.

COE, W. R., 1932. Regeneration in nemerteans. III. Regeneration in *Lineus pictifrons*. *J. exp. Zool.*, **61**, 29–43.

COE, W. R., 1933. Metameric ganglia connected with the dorsal nerve in a nemertean. *Zool. Anz.*, **102**, 237–40.

COE, W. R., 1934a. Regeneration in nemerteans. IV. Cellular changes involved in restitution and reorganisation. *J. exp. Zool.*, **67**, 283–314.

COE, W. R., 1934b. Analysis of the regenerative processes in nemerteans. *Biol. Bull. mar. biol. Lab., Woods Hole*, **66**, 304–15.

COE, W. R., 1934c. New nemerteans from Hawaii. *Occ. Pap. Bernice P. Bishop Mus.*, **10**, 3–9.

COE, W. R., 1935. Bathypelagic nemerteans collected within a 25 mile circle near Bermuda. *Zool. Anz.*, **111**, 315–17.

COE, W. R., 1936. Plankton of the Bermuda Oceanographic Expeditions. VI. Bathypelagic nemerteans taken in the years 1929, 1930 and 1931. *Zoologica, N.Y.*, **21**, 97–113.

COE, W. R., 1938. A new genus and species of Hoplonemertea having differential bipolar sexuality. *Zool. Anz.*, **124**, 220–4.

COE, W. R., 1939. Sexual phases in terrestrial nemerteans. *Biol. Bull. mar. biol. Lab., Woods Hole*, **76**, 416–27.

COE, W. R., 1940a. Notes on the morphology and sexuality of the terrestrial nemertean, *Geonemertes palaensis*. *Occ. Pap. Bernice P. Bishop Mus.*, **15**, 205–11.

COE, W. R., 1940b. Revision of the nemertean fauna of the Pacific coasts of North, Central and northern South America. *Allan Hancock Pacif. Exped.*, **2**, no. 13, 247–323.

COE, W. R., 1943. Biology of the nemerteans of the Atlantic coast of North America. *Trans. Conn. Acad. Arts Sci.*, **35**, 129–328.

COE, W. R., 1944a. A new species of hoplonemertean (*Paranemertes biocellatus*) from the Gulf of Mexico. *J. Wash. Acad. Sci.*, **34**, 407–9.

COE, W. R., 1944b. Nemerteans from the northwest coast of Greenland and other Arctic seas. *J. Wash. Acad. Sci.*, **34**, 59–61.

COE, W. R., 1944c. Geographical distribution of the nemerteans of the Pacific coast of North America, with descriptions of two new species. *J. Wash. Acad. Sci.*, **34**, 27–32.

COE, W. R., 1945a. Plankton of the Bermuda Oceanographic Expeditions. XI. Bathypelagic nemerteans of the Bermuda area and other parts of the north and south Atlantic oceans, with evidence as to their means of dispersal. *Zoologica, N.Y.*, **30**, 145–68.

COE, W. R., 1945b. *Malacobdella minuta*, a new commensal nemertean. *J. Wash. Acad. Sci.*, **35**, 65–7.

COE, W. R., 1947. Nemerteans of the Hawaiian and Marshall Islands. *Occ. Pap. Bernice P. Bishop Mus.*, **19**, 101–6.

COE, W. R., 1950. Nemerteans from Antarctica and the Antarctic Ocean. *J. Wash. Acad. Sci.*, **40**, 56–9.

COE, W. R., 1951a. The nemertean faunas of the Gulf of Mexico and of southern Florida. *Bull. mar. Sci. Gulf Caribb.*, **1**, 149–86.

COE, W. R., 1951b. Geographical distribution of the nemerteans of the northern coast of the Gulf of Mexico as compared with those of the southern coast of Florida, with descriptions of three new species. *J. Wash. Acad. Sci.*, **41**, 328–31.

COE, W. R., 1952. Geographical distribution of the species of nemerteans of the Arctic Ocean near Point Barrow, Alaska. *J. Wash. Acad. Sci.*, **42**, 55–8.

COE, W. R., 1954. The nemertean fauna of the Gulf of Mexico. *Fishery Bull. Fish Wildl. Serv., U.S.*, **55**, 303–9.

COE, W. R., 1959. Nemertea. In: *Freshwater Biology*, 2nd ed. Eds. H. B. Ward and G. C. Whipple, John Wiley, New York, pp. 366–7.

COE, W. R., and BALL, S. C., 1920. The pelagic nemertean *Nectonemertes*. *J. Morph.*, **34**, 457–85.

COE, W. R., and KUNKEL, B. W., 1903. A new species of nemertean (*Cerebratulus melanops*) from the Gulf of St Lawrence. *Biol. Bull. mar. biol. Lab., Woods Hole*, **4**, 119–24.

COLGAN, N., 1916. Observations on phototropism and the development of eyespots in the marine nemertine, *Lineus gesserensis*. *Ir. Nat.*, **25**, 7–12.

CORDERO, E. H., 1943. Hallazgos en diversos países de Sud América de nemertinos de agua dulce del gènero *Prostoma*. *Anais Acad. bras. Cienc.*, **15**, 125–35.

CORRÊA, D. D., 1948. *Ototyphlonemertes* from the Brazilian coast. *Comun. zool. Mus. Hist. nat. Montev.*, **2**, 1–12.

CORRÊA, D. D., 1950. Sôbre *Ototyphlonemertes* do Brasil. *Bolm Fac. Filos. Ciênc. Univ. S Paulo*, **15**, 203–33.

CORRÊA, D. D., 1951. Freshwater nemertines from the Amazon region and from South Africa. *Bolm Fac. Filos. Ciênc. Univ. S Paulo*, **16**, 257–69.

CORRÊA, D. D., 1953a. Sôbre a locomoção e a neurofisiologia de Nemertinos. *Bolm Fac. Filos. Ciênc. Univ. S Paulo*, **18**, 129–47.

CORRÊA, D. D., 1953b. Sôbre a neurofisiologia locomotora de Hoplonemertinos e a taxonomia de *Ototyphlonemertes*. *Anais Acad. bras. Cienc.*, **25**, 545–55.

CORRÊA, D. D., 1954. Nemertinos do litoral Brasileiro. *Concur. Docên.-Livr. Cad. Zool. Fac. Filos. Ciênc. Univ. S Paulo*, 1–91.

CORRÊA, D. D., 1956. Estudo de Nemertinos Mediterrâneos (Palaeo e Hetero-
    nemertini). *Anais Acad. bras. Cienc.*, **28**, 195–214.
CORRÊA, D. D., 1957. Nemertinos do litoral Brasileiro – VI. *Anais Acad. bras.
    Cienc.*, **29**, 251–71.
CORRÊA, D. D., 1958. Nemertinos do litoral Brasileiro (VII). *Anais Acad. bras.
    Cienc.*, **29**, 441–55.
CORRÊA, D. D., 1961. Nemerteans from Florida and Virgin Islands. *Bull. mar.
    Sci. Gulf Caribb.*, **11**, 1–44.
CORRÊA, D. D., 1963. Nemerteans from Curaçao. *Stud. Fauna Curaçao*, **17**,
    41–56.
CORRÊA, D. D., 1964. Nemerteans from California and Oregon. *Proc. Calif.
    Acad. Sci.*, Ser. 4, **31**, 515–58.
CORRÊA, D. D., 1966. A new hermaphroditic nemertean. *Anais Acad. bras.
    Cienc.*, **38**, 365–9.
COWEY, J. B., 1952. The structure and function of the basement membrane
    muscle system in *Amphiporus lactifloreus* (Nemertea). *Q. Jl. microsc. Sci.*, **93**,
    1–15.
CRAVENS, M. R., and HEATH, H., 1906. The anatomy of a new species of *Necto-
    nemertes*. *Zool. Jb.*, **23**, 337–56.
CROZIER, W. J., 1917. Note on the habitat of *Geonemertes agricola*. *Am. Nat.*,
    51, 758–60.
CUVIER, G., 1817. Le règne animal distribué d'après son organisation pour servir
    de base a l'histoire naturelle des animaux, et d'introduction a l'anatomie
    comparée. Vol. 4. Les Zoophytes ou animaux rayonnes. Fortin, Masson et
    Cie., Paris, 160 pp.
DAKIN, W. J., and FORDHAM, M. G. C., 1931. A new and peculiar marine
    nemertean from the Australian coast. *Nature, Lond.*, **128**, 796.
DAKIN, W. J., and FORDHAM, M. G. C., 1936. The anatomy and systematic
    position of *Gorgonorhynchus repens*, gen. n., sp. n.: a new genus of nemer-
    tines characterised by a multi-branched proboscis. *Proc. zool. Soc. Lond.*,
    461–83.
DANIELLI, J. F., and PANTIN, C. F. A., 1950. Alkaline phosphatase in proto-
    nephridia of terrestrial nemertines and planarians. *Q. Jl. microsc. Sci.*, **91**,
    209–13.
DAVIS, C. C., 1965. A study of the hatching process in aquatic invertebrates. XX.
    The Blue Crab, *Callinectes sapidus*, Rathbun, XXI. The nemertean, *Carcino-
    nemertes carcinophila* (Kölliker). *Chesapeake Sci.*, **6**, 201–8.
DAWYDOFF, C., 1910. Restitution von Kopfstücken, die vor der Mundöffnung
    abgeschnitten waren, bei den Nemertinen (*Lineus lacteus*). *Zool. Anz.*, **36**,
    1–6.
DAWYDOFF, C., 1928. Sur l'embryologie des Protonémertes. *C. r. hebd. Séanc.
    Acad. Sci., Paris*, **186**, 531–3.
DAWYDOFF, C., 1940. Les formes larvaires de polyclades et de némertes du
    plancton indochinois. *Bull. biol. Fr. Belg.*, **74**, 443–96.
DAWYDOFF, C., 1952. Contribution à l'étude des invertébrés de la faune marine
    benthique de l'Indochine. *Bull. biol. Fr. Belg.*, Suppl. no. 37, 1–158.
* DEARBORN, J. H., 1965. *Ecological and faunistic investigations of the marine
    benthos at McMurdo Sound, Antarctica.* Ph.D. thesis, University of Stanford.
DEBAISIEUX, P., 1920. Haplosporidium (*Minchinia*) chitonis Lank., *Haplospori-
    dium nemertis*, nov. sp., et le groupe des Haplosporidies. *Cellule*, **30**, 293–311.
* DELLE CHIAJE, S., 1825. Memorie sulla storia e notomia degli animali senze
    vertebre del regno di Napoli. Naples, **2**, 406–27.

DELLE CHIAJE, S., 1841. Descrizione et notomia degli animali invertebrati della Sicilia citeriore. Naples, 3, 125–30.

DELSMAN, H. C., 1915. Eifurchung und Gastrulation bei *Emplectonema gracile* Stimpson. *Tijdschr. ned. dierk. Vereen.*, Ser. 2, 14, 68–114.

DENDY, A., 1892. On an Australian land nemertine (*Geonemertes australiensis*, n. sp). *Proc. R. Soc. Vict.*, 4 (New Ser.), 85–122.

DENDY, A., 1893. Notes on the mode of reproduction of *Geonemertes australiensis*. *Proc. R. Soc. Vict.*, 5 (New Ser.), 127–30.

DESOR, E., 1848. On the embryology of *Nemertes*, with an appendix on the embryonic development of *Polynöe*; and remarks upon the embryology of marine worms in general. *Boston J. nat. Hist.*, 6, 1–18.

DEWOLETZKY, R., 1887. Das Seitenorgan der Nemertinen. *Arb. zool. Inst. Univ. Wien*, 7, 233–80.

DU BOIS-REYMOND MARCUS, E., 1948. An Amazonian heteronemertine. *Bolm Fac. Filos. Ciênc. Univ. S Paulo*, 13, 93–109.

DUGÈS, A., 1830. Aperçu de quelques observations nouvelle sur les planaires et plusieurs genres voisins. *Annls Sci. nat.*, Ser. 1, 21, 72–90.

DU PLESSIS, G., 1893. Organisation et genre de vie de l'*Emea lacustris*, nemertien des environs de Genève. *Revue suisse Zool.*, 1, 329–57.

EGGERS, F., 1924. Zur Bewegungsphysiologie der Nemertinen. I. *Emplectonema*. *Z. vergl. Physiol.*, 1, 579–89.

EGGERS, F., 1935. Zur Bewegungsphysiologie von *Malacobdella grossa* Müll. *Z. wiss. Zool.*, 147, 101–31.

* EHRENBERG, C. G., 1831. Symbolae physicae seu icones et descriptiones corporum naturalium novorum aut minus cognitorum. Phytozoa turbellaria. *Abh. dt. Akad. Wiss. Berl.*

FABRICIUS, O., 1780. *Fauna Groenlandica*. Hafniae et Lipsiae, 452 pp.

FABRICIUS, O., 1797. Beskrivelse over 4 livet bekjendte Flad-Orme (*Planaria angulata, fuscescens, candida, brunnea*). *Skrift. nat. Selsk. Kjöbenhavn*, 4, 52–66.

FENCHEL, T., 1965. Ciliates from Scandinavian molluscs. *Ophelia*, 2, 71–3.

FISHER, F. M., and CRAMER, N. M., 1967. New observations on the feeding mechanism in *Lineus ruber* (Rhynchocoela). *Biol. Bull. mar. biol. Lab., Woods Hole*, 132, 464.

FOSHAY, E. A., 1912. *Nectonemertes japonica*, a new nemertean. *Zool. Anz.*, 40, 50–3.

FOX, D. L., 1954. An acidogenic carotenoid in a bathypelagic nemertean worm. *Nature, Lond.*, 173, 583.

FRIEDRICH, H., 1935. Studien zur Morphologie, Systematik und Ökologie der Nemertinen der Kieler Bucht. *Arch. Naturg.*, 4, 293–375.

FRIEDRICH, H., 1936. Einige Bemerkungen zur Anatomie von *Tubulanus borealis* n. sp., einer neuen Paläonemertine aus der Nordsee. *Zool. Anz.*, 116, 101–8.

FRIEDRICH, H., 1940. Nemertini. In: The fishery grounds near Alexandria. *Fuad I. Inst. Hydrobiol. Fish., Notes Mem.*, no. 38, 6 pp.

FRIEDRICH, H., 1955. Beiträge zu einer Synopsis der Gattungen der Nemertini monostilifera nebst Bestimmungsschlüssel. *Z. wiss. Zool.*, 158, 133–92.

FRIEDRICH, H., 1956. Zur Morphologie des Vorderdarmes der monostiliferen Hoplonemertinen. *Veröff. Inst. Meeresforsch. Bremerh.*, 4, 45–53.

FRIEDRICH, H., 1957. Beiträge zur Kenntnis der arktischen Hoplonemertinen. *Vidensk. Meddr. dansk naturh. Foren.*, 119, 129–54.

FRIEDRICH, H., 1958. Nemertini. *Zoology Iceland*, 2, Pt. 18, 1–24.

FRIEDRICH, H., 1960. Bemerkungen über die Gattung *Micrura* Ehrenberg 1831 und zur Klassifikation der Heteronemertinen nebst vorläufigem Bestimmungsschlüssel. *Veröff. Inst. Meeresforsch. Bremerh.*, **7**, 48–62.

FRIEDRICH, H., 1965. Gesamtverzeichnis der Literatur über die Nemertinen. *Veröff. Überseemus. Bremen.* **3**, 204–44,

FRIEDRICH, H., 1969. Ergänzungen zum Gesamtverzeichnis der Literatur über die Nemertinen I. *Veröff. Überseemus. Bremen*, **4**, 9–16.

FRIEDRICH, H., 1970. Nemertinen aus Chile. *Sarsia*, **40**, 1–80.

GIBSON, R., 1967. Occurrence of the entocommensal rhynchocoelan, *Malacobdella grossa*, in the Oval Piddock, *Zirfaea crispata*, on the Yorkshire coast. *J. mar. biol. Ass. U.K.*, **47**, 301–17.

GIBSON, R., 1968. Studies on the biology of the entocommensal rhynchocoelan *Malacobdella grossa*. *J. mar. biol. Ass. U.K.*, **48**, 637–56.

GIBSON, R., 1970. The nutrition of *Paranemertes peregrina* (Rhynchocoela: Hoplonemertea). II. Observations on the structure of the gut and proboscis, site and sequence of digestion, and food reserves. *Biol. Bull. mar. biol. Lab.*, *Woods Hole*, **139**, 92–106.

GIBSON, R., and JENNINGS, J. B., 1967. 'Leucine aminopeptidase' activity in the blood system of rhynchocoelan worms. *Comp. Biochem. Physiol.*, **23**, 645–51.

GIBSON, R., and JENNINGS, J. B., 1969. Observations on the diet, feeding mechanisms, digestion and food reserves of the entocommensal rhynchocoelan *Malacobdella grossa*. *J. mar. biol. Ass. U.K.*, **49**, 17–32.

GIBSON, R., and MOORE, J., 1971. Occurrence of the freshwater hoplonemertean *Prostoma graecense* (Böhmig) in Tasmania. *Freshwat. Biol.*, **1**, 193–5.

GIBSON, R., and YOUNG, J. O., 1971. *Prostoma jenningsi* sp. nov., a new British freshwater hoplonemertean. *Freshwat. Biol.*, **1**, 121–7.

GIRARD, C., 1851. An essay on the classification of *Nemertes* and Planariae: preceded by some general considerations on the primary divisions of the animal kingdom. *Am. J. Sci.*, Ser. 2, **11**, 41–53.

GIRARD, C., 1853. Descriptions of new nemerteans and planarians from the coast of the Carolinas. *Proc. Acad. nat. Sci. Philad.*, **6**, 365–7.

GONTCHAROFF, M., 1948. Note sur l'alimentation de quelques Némertes. *Annls Sci. nat., Zool.*, Ser. 11, **10**, 75–8.

GONTCHAROFF, M., 1949. Sur un cas de régénération anormale chez la Némerte *Lineus ruber β*. *C. r. hebd. Séanc. Acad. Sci., Paris*, **228**, 1757–8.

GONTCHAROFF, M., 1951. Biologie de la régénération et de la reproduction chez quelques Lineidae de France. *Annls Sci. nat., Zool.*, Ser. 11, **13**, 149–235.

GONTCHAROFF, M., 1952. Réactions à la lumière de *Lineus ruber* (Némertien) en éclairage ventral. *C. r. hebd. Séanc. Acad. Sci., Paris*, **235**, 1690–2.

GONTCHAROFF, M., 1953. Le phototropisme chez *Lineus ruber* et *Lineus sanguineus* au cours de la régénération des yeux. *Annls Sci. nat., Zool.*, Ser. 11, **15**, 369–72.

GONTCHAROFF, M., 1955. Nemertes. Inventaire de la Faune Marine de Roscoff, Suppl. 7, 3–15.

GONTCHAROFF, M., 1957. Étude des rhabdites de la trompe de *Lineus ruber* (Némertien) au microscope électronique. *C. r. hebd. Séanc. Acad. Sci., Paris*, **244**, 1539–41.

GONTCHAROFF, M., 1958a. L'autotomie spontanée de la trompe chez *Eunemertes echinoderma*. *C. r. hebd. Séanc. Acad. Sci., Paris*, **246**, 1296–7.

GONTCHAROFF, M., 1958b. L'autogreffe de la trompe chez *Eunemertes echinoderma* (Marion). *C. r. hebd. Séanc. Acad. Sci., Paris*, **247**, 246–7.

GONTCHAROFF, M., 1959. Rearing of certain nemerteans (genus *Lineus*). *Ann. N.Y. Acad. Sci.*, **77**, 93–5.

GONTCHAROFF, M., 1960. Le développement post-embryonnaire et la croissance chez *Lineus ruber* et *Lineus viridis* (Nemertes Lineidae). *Annls Sci. nat., Zool.*, Ser. 12, **2**, 225–79.

GONTCHAROFF, M., and BIERNE, J., 1962. Sur l'existence des cellules basophiles chez *Lineus ruber*. *C. r. hebd. Séanc. Acad. Sci., Paris*, **255**, 570–1.

GONTCHAROFF, M., and DEROUX, G., 1949. Sur une némerte d'eau douce (*Prostoma eilhardi* Montg.) récoltée à la Baiorry (Pyr. Orient.). *Bull. Soc. zool. Fr.*, **74**, 133–6.

GONTCHAROFF, M., and LECHENAULT, H., 1966. Ultrastructure et histochemie des glandes sous épidermiques chez *Lineus ruber* et *Lineus viridis*. *Histochemie*, **6**, 320–35.

GRAY, J., 1940. Aspects of animal locomotion. *Proc. R. Soc., B*, **128**, 28–62.

GREEN, J., 1968. *The Biology of Estuarine Animals*. Sidgwick and Jackson, London, 401 pp. (Nemertea 75–6, 130–1).

HAMMARSTEN, O., 1918. Beitrag zur Embryonalentwicklung der *Malacobdella grossa*. *Arb. zool. Inst. Stock.*, **11**, 195.

HENLEY, C., 1970. Changes in microtubules of cilia and flagella following negative staining with phosphotungstic acid. *Biol. Bull. mar. biol. Lab., Woods Hole*, **139**, 265–76.

HICKMAN, V. V., 1963. The occurrence in Tasmania of the land nemertine, *Geonemertes australiensis* Dendy, with some account of its distribution, habits, variations and development. *Pap. Proc. R. Soc. Tasm.*, **97**, 63–75.

HOFFMANN, C. K., 1877. Zur Anatomie und Ontogenie von *Malacobdella*. *Niederl. Arch. Zool.*, **4**, 1–27.

HÖRSTADIUS, S., 1937. Experiments on determination in the early development of *Cerebratulus lacteus*. *Biol. Bull. mar. biol. Lab., Woods Hole*, **73**, 317–42.

HUBRECHT, A. A. W., 1875. Some remarks about the minute anatomy of Mediterranean nemerteans. *Q. Jl. microsc. Sci.*, **15**, 249–56.

HUBRECHT, A. A. W., 1880a. The peripheral nervous system in Palaeo- and Schizonemertini, one of the layers of the body wall. *Q. Jl. microsc. Sci.*, **20**, 431–42.

HUBRECHT, A. A. W., 1880b. Zur Anatomie und Physiologie des Nervensystems der Nemertinen. *Verh. K. Akad. Wet.*, **20**, 47 pp.

HUBRECHT, A. A. W., 1883. On the ancestral form of the Chordata. *Q. Jl. microsc. Sci.*, **23**, 349–68.

HUBRECHT, A. A. W., 1886. Contributions to the embryology of the Nemertea. *Q. Jl. microsc. Sci.*, **26**, 417–48.

HUBRECHT, A. A. W., 1887a. Report on the Nemertea collected by H.M.S. *Challenger* during the years 1873–6. *Rep. sci. Res. Voy. H.M.S. Challenger, Zool.*, **19**, 1–147.

HUBRECHT, A. A. W., 1887b. The relation of the Nemertea to the Vertebrata. *Q. Jl. microsc. Sci.*, **27**, 605–44.

HUMES, A. G., 1941. The male reproductive system in the nemertean genus *Carcinonemertes*. *J. Morph.*, **69**, 443–54.

HUMES, A. G., 1942. The morphology, taxonomy, and bionomics of the nemertean genus *Carcinonemertes*. *Illinois biol. Monogr.*, **18**, 1–105.

HUSCHKE, E., 1830. Beschreibung und Anatomie eines an Sicilien gefundenen Meerwurms, *Notospermus drepanensis*. *Oken's Isis*, **23**, 682–3.

HUXLEY, T. H., 1877. *A Manual of the Anatomy of Invertebrated Animals*.

Chapter 4. The Turbellaria, the Rotifera, the Trematoda, and the Cestoidea. pp. 176 et seq.

HYLBOM, R., 1957. Studies on palaeonemerteans of the Gullmar Fiord area (west coast of Sweden). *Ark. Zool.*, **10**, 539–82.

HYMAN, L. H., 1951. *The Invertebrates, Vol. II: Platyhelminthes and Rhynchocoela.* McGraw-Hill Book Co. Inc., New York, 550 pp.

IWATA, F., 1951. Nemerteans in the vicinity of Onomichi. *J. Fac. Sci. Hokkaido Univ.*, Ser. 6, Zool., **10**, 135–8.

IWATA, F., 1952. Nemertini from the coasts of Kyusyu. *J. Fac. Sci. Hokkaido Univ.*, Ser. 6, Zool., **11**, 126–48.

IWATA, F., 1954a. The fauna of Akkeshi Bay. XX. Nemertini in Hokkaido (revised report). *J. Fac. Sci. Hokkaido Univ.*, Ser. 6, Zool., **12**, 1–39.

IWATA, F., 1954b. Invertebrate fauna of the intertidal zone of the Tokara Islands. X. Nemertini. *Publs Seto mar. biol. Lab.*, **4**, 27–31.

IWATA, F., 1954c. Some nemerteans from the coasts of the Kii Peninsula. *Publs Seto mar. biol. Lab.*, **4**, 33–42.

IWATA, F., 1957a. On the early development of the nemertine *Lineus torquatus* Coe. *J. Fac. Sci. Hokkaido Univ.*, Ser. 6, Zool., **13**, 54–8.

IWATA, F., 1957b. Nemerteans from Sagami Bay. *Publs Akkeshi mar. biol. Stn*, no. 7, 1–31.

IWATA, F., 1958. On the development of the nemertean *Micrura akkeshiensis. Embryologia*, **4**, 103–31.

IWATA, F., 1960a. Studies on the comparative embryology of nemerteans with special reference to their interrelationships. *Publs Akkeshi mar. biol. Stn*, no. 10, 1–51.

IWATA, F., 1960b. The life history of the Nemertea. *Bull. biol. Stn Asamushi*, **10**, 95–7.

IWATA, F., 1967. *Uchidana parasita* nov. gen. et nov. sp., a new parasitic nemertean from Japan with peculiar morphological characters. *Zool. Anz.*, **178**, 122–36.

IWATA, F., 1968. On the circulatory and nephridial organs of Nemertinea and their osmoregulatory function. *Proc. Jap. Soc. Syst. Zool.*, no. 4, 7–9.

IWATA, F., 1970. On the brackish-water nemerteans from Japan, provided with special circulatory and nephridial organs useful for osmoregulation. *Zool. Anz.*, **184**, 133–54.

JACKSON, L. H., 1935. Sense organs in *Malacobdella. Nature, Lond.*, **135**, 792.

JENKINS, O. P., and CARLSON, A. J., 1903. The rate of the nervous impulse in the ventral nerve-cord of certain worms. *J. comp. Neurol.*, **13**, 259–89.

JENNINGS, J. B., 1960. Observations on the nutrition of the rhynchocoelan *Lineus ruber* (O. F. Müller). *Biol. Bull. mar. biol. Lab., Woods Hole*, **119**, 189–96.

JENNINGS, J. B., 1962. A histochemical study of digestion and digestive enzymes in the rhynchocoelan *Lineus ruber* (O. F. Müller). *Biol. Bull. mar. biol. Lab., Woods Hole*, **122**, 63–72.

JENNINGS, J. B., 1968. A new astomatous ciliate from the entocommensal rhynchocoelan *Malacobdella grossa* (O. F. Müller). *Arch. Protistenk.*, **110**, 422–5.

JENNINGS, J. B., 1969. Ultrastructural observations on the phagocytic uptake of food materials by the ciliated cells of the rhynchocoelan intestine. *Biol. Bull. mar. biol. Lab., Woods Hole*, **137**, 476–85.

JENNINGS, J. B., and GIBSON, R., 1968. The structure and life history of *Haplosporidium malacobdellae* sp. nov., a new sporozoan from the entocommensal rhynchocoelan *Malacobdella grossa* (O. F. Müller). *Arch. Protistenk.*, **111**, 31–7.

JENNINGS, J. B., and GIBSON, R., 1969. Observations on the nutrition of seven species of rhynchocoelan worms. *Biol. Bull. mar. biol. Lab., Woods Hole*, **136**, 405–33.

JENSEN, D. D., 1960. Hoplonemertines, myxinoids and Deuterostome origins. *Nature, Lond.*, **188**, 649–50.

JENSEN, D. D., 1963. Hoplonemertines, myxinoids, and vertebrate origins. In: *The Lower Metazoa, Comparative Biology and Phylogeny*. Ed. E. C. Dougherty, Univ. Calif. Press, pp. 113–26.

JOHNSTON, G., 1837. Miscellanea Zoologica. II. A description of some planarian worms. *Mag. Zool. Bot.*, **1**, 529–38.

JOHNSTON, G., 1865. *A Catalogue of the British Non-Parasitical Worms in the Collection of the British Museum*. Taylor and Francis, London, 365 pp.

JOUBIN, L., 1914. Sur deux cas d'incubation chez des Némertiens antarctiques. *C. r. hebd. Séanc. Acad. Sci., Paris*, **158**, 430–2.

KADIS, S., 1951. The effects of inanition on the freshwater nemertean worm *Prostoma rubrum*. *Proc. Va. Acad. Sci.*, **2**, 314.

KAMEMOTO, F. I., 1957. Cholinesterase in the nemertean *Prostoma rubrum*. *Science, N.Y.*, **125**, 351–2.

KANDA, S., 1939. The luminescence of a nemertean, *Emplectonema kandai*, Kato. *Biol. Bull. mar. biol. Lab., Woods Hole*, **77**, 166–73.

KARLING, T. G., 1933. Ein Beitrag zur Kenntnis der Nemertinen des Finnischen Meerbusens. *Memo. Soc. Fauna Flora fenn.*, **10**, 76–90.

KARLING, T. G., 1965. *Haplopharynx rostratus* Meixner (Turbellaria) mit den Nemertinen verglichen. *Z. zool. Syst. Evolut.*, **3**, 1–18.

KATO, K., 1939. A new luminous species of the Nemertea, *Emplectonema kandai* sp. nov. *Jap. J. Zool.*, **8**, 251–4.

KEM, W. R., 1971. A study of the occurrence of anabaseine in *Paranemertes* and other nemertines. *Toxicon*, **9**, 23–32.

KEM, W. R., ABBOTT, B. C., and COATES, R. M., 1971. Isolation and structure of a hoplonemertine toxin. *Toxicon*, **9**, 15–22.

KING, H., 1939. Amphiporine, an active base from the marine worm *Amphiporus lactifloreus*. *J. chem. Soc.*, Pt. 2, 1365–6.

KIPKE, S., 1932. Studien über Regenerationserscheinungen bei Nemertinen (*Prostoma graecense* Böhmig). *Zool. Jb.*, **51**, 1–66.

KIRSTEUER, E., 1963a. Beitrag zur Kenntnis der Systematik und Anatomie der adriatischen Nemertinen (Genera *Tetrastemma, Oerstedia, Oerstediella*). *Zool. Jb.*, **80**, 555–616.

KIRSTEUER, E., 1963b. Zur Ökologie systematischer Einheiten bei Nemertinen. *Zool. Anz.*, **170**, 343–54.

KIRSTEUER, E., 1965. Über das Vorkommen von Nemertinen in einem tropischen Korallenriff. 4. Hoplonemertini monostilifera. *Zool. Jb.*, **92**, 289–326.

KIRSTEUER, E., 1966. Über *Carcinonemertes carcinophila* (Kölliker) aus der Nordadria. *Zool. Anz.*, **176**, 205–12.

KIRSTEUER, E., 1967a. New marine nemerteans from Nossi Be, Madagascar. *Zool. Anz.*, **178**, 110–22.

KIRSTEUER, E., 1967b. Marine, benthonic nemerteans: how to collect and preserve them. *Am. Mus. Novit.*, no. 2290, 1–10.

KNIGHT-JONES, E. W., 1954. Relations between metachronism and the direction of ciliary beat in Metazoa. *Q. Jl. microsc. Sci.*, **95**, 503–21.

208   Bibliography

KÖLLIKER, A., 1848. Beiträge zur Kenntniss niederer Thiere. *Z. wiss. Zool.*, **1**, 1–37.
KOSTANECKI, M. C., 1902a. Dojrzewanie i zapłodnienie jajka u *Cerebratulus marginatus*. *Bull. int. Acad. Sci. Lett. Cracovie*, no. 5, 270–7.
KOSTANECKI, M. C., 1902b. Nieprawidłowe figury mitotyczne przy wydzielaniu ciałek kierunkowych w jajkach *Cerebratulus marginatus*. *Bull. int. Acad. Sci. Lett. Cracovie*, no. 5, 278–89.
LAMARCK, J. B., 1801. *Système des Animaux sans Vertèbres*. Deterville, Paris, 432 pp.
LANKESTER, E. R., 1872. A contribution to the knowledge of haemoglobin. *Proc. R. Soc.*, **21**, 70–81.
LEBEDINSKY, J., 1896. Zur Entwicklungsgeschichte der Nemertinen. *Biol. Zbl.*, **16**, 577–86.
LEBEDINSKY, J., 1897. Zur Entwicklungsgeschichte der Nemertinen. *Biol. Zbl.*, **17**, 113–24.
LECHENAULT, H., 1962. Sur l'existence de cellules neurosécrétrices dans les ganglions cérébroïdes des Lineidae (Hétéronémertes). *C. r. hebd. Séanc. Acad. Sci., Paris*, **255**, 194–6.
LECHENAULT, H., 1963. Sur l'existence de cellules neurosécrétrices chez les Hoplonémertes. Caractéristiques histochimiques de la neurosécrétion chez les Némertes. *C. r. hebd. Séanc. Acad. Sci., Paris*, **256**, 3201–3.
LECHENAULT, H., 1965. Neurosécrétion et osmorégulation chez les Lineidae (Hétéronémertes). *C. r. hebd. Séanc. Acad. Sci., Paris*, **261**, 4868–71.
LEIDY, J., 1850. Description of new genera of Vermes. *Proc. Acad. nat. Sci. Philad.*, **5**, 124–6.
LEIDY, J., 1851. Corrections and additions to former papers on helminthology published in the Proceedings of the Academy. *Proc. Acad. nat. Sci. Philad.*, **5**, 284–90.
LEPORI, N. G., 1949. Rinvenimento di un Nemerteo (*Prostoma graecense*) (Böhmig), nelle acque dolci dei dintorni di Pisa. *Boll. Zool.*, **16**, 33–40.
LEPPÄKOSKI, E., 1969. Transitory return of the benthic fauna of the Bornholm Basin after extermination by oxygen insufficiency. *Cah. Biol. mar.*, **10**, 163–72.
LEUCKART, F. S., 1830. Notiz über *Meckelia somatotomus*. *Oken's Isis*, **23**, 575–6.
LING, E.-A., 1969a. The structure and function of the cephalic organ of a nemertine *Lineus ruber*. *Tissue Cell*, **1**, 503–24.
LING, E.-A., 1969b. The structure and function of the cephalic organs of the nemertines (*Lineus ruber* and *Amphiporus lactifloreus*). Ph.D. thesis, Univ. of Cambridge, 202 pp.
LING, E.-A., 1970. Further investigations on the structure and function of cephalic organs of a nemertine *Lineus ruber*. *Tissue Cell*, **2**, 569–88.
LING, E.-A., 1971. The proboscis apparatus of the nemertine *Lineus ruber*. *Phil. Trans. R. Soc., B*, **262**, 1–22.
LINNAEUS, C., 1788. *Systema Naturae*. Editio decima tertia, aucta, reformata. Ed. Gmelin, Leipsiae, **1**, Pt. 6, Vermes Testacea, 3087–94, 3098.
LÖNNBERG, E., 1933. Zur Kenntniss der Carotinoide bei marinen Evertebraten. *Ark. Zool.*, **25A**, no. 1, 1–17.
LÖNNBERG, E., 1934. Weitere Beiträge zur Kenntnis der Carotinoide der marinen Evertebraten. *Ark. Zool.*, **26A**, no. 7, 1–36.
MCCAUL, W. E., 1963. Rhynchocoela: nemerteans from marine and estuarine waters of Virginia. *J. Elisha Mitchell scient. Soc.*, **79**, 111–24.

Bibliography

McDermott, J. J., 1966. The biology of a nemertean parasite of pinnotherid crabs. *Am. Zoologist*, **6**, 166.

Macfarlane, J. M., 1918. *The Causes and Course of Organic Evolution.* Macmillan, New York, 875 pp.

McIntosh, W. C., 1868. On the boring of certain annelids. *Ann. Mag. nat. Hist.*, Ser. 4, **2**, 276–95.

McIntosh, W. C., 1873–4. A Monograph of the British Annelids. Pt. 1. The Nemerteans. *Ray Soc. Publs*, 214 pp.

McIntosh, W. C., 1876. On the central nervous system, the cephalic sacs, and other points in the anatomy of the Lineidae. *J. Anat. Physiol., Lond.*, **10**, 231–52.

McIntosh, W. C., 1906. Notes from the Gatty Marine Laboratory, St Andrews, no. XXVII, Pt. 3. On bifid nemerteans (*Cerebratulus angulatus*, O.F.M., = *marginatus*, Renier?) from Aberdeen and Naples. *Ann. Mag. nat. Hist.*, Ser. 7, **17**, 74–8.

Maclaren, N. H. W., 1901. On the blood vascular system of *Malacobdella grossa*. *Zool. Anz.*, **24**, 126–9.

MacLeay, W. S., 1839. Note on the Annelida. In: Sir R. I. Murchison's *The Silurian System*, Pt. 1, John Murray, London, 768 pp.

Marion, A. F., 1874. Recherches sur les animaux inférieurs du Golfe de Marseille. Troisième article. Remarques complémentaires sur le *Borlasia kefersteinii*. *Annls Sci. nat., Zool.*, Ser. 6, **6**, no. 1, 19–30.

Mayr, E., Linsley, E. G., and Usinger, R. L., 1953. *Methods and Principles of Systematic Zoology*. McGraw-Hill Book Co. Inc., New York, 336 pp.

Mendes, M. V., 1949. Respiration in worms. *Anais Acad. bras. Cienc.*, **21**, 19–54.

Metschnikoff, E., 1869. Studien über die Entwickelung der Echinodermen und Nemertinen. *Zap. imp. Akad. Nauk*, Ser. 7, **14**, no. 8, 1–73.

Monastero, S., 1928. Esperienze sulla rigenerazione dei Nemertini (*Lineus nigricans* Bürger 1892 e *Prostoma melanocephalum melan.* Johnst. 1837). *Boll. Ist. zool. R. Univ. Palermo*, **2**, 1–16.

Montgomery, T. H., 1897. Studies on the elements of the central nervous system of the Heteronemertini. *J. Morph.*, **13**, 381–444.

Moore, J., and Gibson, R., 1972. On a new genus of freshwater hoplonemertean from Campbell Island. *Freshwater Biol.*, **2** (in press).

Moore, J., and Gibson, R., 1973. A new genus of freshwater hoplonemertean from New Zealand. *Zool. J. Linn. Soc.*, **52** (in press).

Morse, M., 1912. Artificial parthenogenesis and hybridisation in the eggs of certain invertebrates. *J. exp. Zool.*, **13**, 471–96.

Müller, G. J., 1962. Contribuţii la studiul nemerţienilor din Marea Neagră (Litoralul Romînesc). *Studii Cerc. Biol., Ser. Biol. Anim.*, **14**, 371–84.

Müller, G. J., 1966. Analiza zoogeografică a faunei de Nemerţieni din Marea Neagră. *Hidrobiologia*, **7**, 131–40.

Müller, J., 1847. Fortsetzung des Berichts über einige neue Thierformen der Nordsee. Pilidium gyrans. *Arch. Anat. Physiol.*, 159–60.

Müller, O. F., 1773. Vermium terrestrium et fluviatilium, seu animalium infusoriorum, helminthicorum et testaceorum, non marinorum, succincta historia. Heineck et Faber, Havniae et Lipsiae, **1**, Pt. 1, 135 pp.

Müller, O. F., 1774. Vermium terrestrium et fluviatilium, seu animalium infusoriorum, helminthicorum et testaceorum, non marinorum, succincta historia. Heineck et Faber, Havniae et Lipsiae, **1**, Pt. 2, 72 pp.

Müller, O. F., 1788–1806. Zoologia Danica seu Animalium Daniae et Norve-

giae Rariorum ac minus notorum descriptiones et historia. Havniae, vols. 1–4.
NAWITZKI, W., 1931. *Procarinina remanei*, Eine neue Paläonemertine der Kieler Förde. *Zool. Jb.*, **54**, 159–234.
NUSBAUM, J., and OXNER, M., 1910. Studien über die Regeneration der Nemertinen, I. Regeneration bei *Lineus ruber* (Müll.). *Arch. EntwMech. Org.*, **30**, 74–132.
NUSBAUM, J., and OXNER, M., 1911a. Die Bildung des ganzen neuen Darmkanals durch Wanderzellen mesodermalen Ursprunges bei der Kopfrestitution des *Lineus lacteus* (Grube) (Nemertine). *Zool. Anz.*, **37**, 302–15.
NUSBAUM, J., and OXNER, M., 1911b. Weitere Studien über die Regeneration der Nemertinen. I. Regeneration bei *Lineus ruber* Müll. *Arch. EntwMech. Org.*, **32**, 349–96.
NUSBAUM, J., and OXNER, M., 1912. Fortgesetzte Studien über Regeneration der Nemertinen. II. Regeneration des *Lineus lacteus* Rathke. *Arch. Entw-Mech. Org.*, **35**, 236–308.
NUSBAUM, J., and OXNER, M., 1913. Die Embryonalentwicklung des *Lineus ruber* Müll. Ein Beitrag zur Entwicklungsgeschichte der Nemertinen. *Z. wiss. Zool.*, **107**, 78–197.
OHUYE, T., 1942. On the blood corpuscles and the hemopoiesis of a nemertean, *Lineus fuscoviridis*, and of a sipunculus, *Dendrostoma minor*. *Sci. Rep. Tôhoku Univ.*, Ser. 4, Biol., **17**, 187–96.
ÖRSTED, A. S., 1844. Entwurf einer Systematischen Einteilung und speciellen Beschreibung der Plattwürmer. Kopenhagen, pp. 76 et seq.
OUCHAKOFF, P., 1925. Seasonal changes on the shore of Kola Bay. *Trav. Soc. Nat. Leningrad*, **54**, 47–71 (in Russian).
OUDEMANS, A. C., 1885. The circulatory and nephridial apparatus of the Nemertea. *Q. Jl. microsc. Sci.*, **25**, Suppl., 1–80.
PALLAS, P. S., 1766. *Miscellanea Zoologica*. Petrum van Cleef, Hagae, 224 pp.
PANTIN, C. F. A., 1947. The nephridia of *Geonemertes dendyi*. *Q. Jl. microsc. Sci.*, Ser. 3, **88**, 15–25.
PANTIN, C. F. A., 1950. Locomotion in British terrestrial nemertines and planarians: with a discussion on the identity of *Rhynchodemus bilineatus* (Mecznikow) in Britain, and on the name *Fasciola terrestris* O. F. Müller. *Proc. Linn. Soc. Lond.*, **162**, 23–37.
PANTIN, C. F. A., 1961a. *Acteonemertes bathamae*, gen. et sp. nov., an upper littoral nemertine from Portobello, New Zealand. *Proc. Linn. Soc. Lond.*, **172**, 153–6.
PANTIN, C. F. A., 1961b. *Geonemertes*: a study in island life. *Proc. Linn. Soc. Lond.*, **172**, 137–52.
PANTIN, C. F. A., 1969. The genus *Geonemertes*. *Bull. Br. Mus. nat. Hist.*, (*Zool.*), **18**, 263–310.
PEARSE, A. S., 1949. Observations on flatworms and nemerteans collected at Beaufort, N.C. *Proc. U.S. natn. Mus.*, **100**, 25–38.
PEDERSEN, K. J., 1968. Some morphological and histochemical aspects of nemertean connective tissue. *Z. Zellforsch. mikrosk. Anat.*, **90**, 570–95.
POLUHOWICH, J. J., 1968. Notes on the freshwater nemertean *Prostoma rubrum*. *Turtox News*, **46**, 2–7.
POLUHOWICH, J. J., 1970. Oxygen consumption and the respiratory pigment in the freshwater nemertean *Prostoma rubrum*. *Comp. Biochem. Physiol.*, **36**, 817–21.
PUNNETT, R. C., 1900. On a collection of nemerteans from Singapore. *Q. Jl. microsc. Sci.*, **44**, 111–39.

PUNNETT, R. C., 1901a. On two new British nemerteans. *Q. Jl. microsc. Sci.*, **44**, 547–64.

PUNNETT, R. C., 1901b. Nemerteans. In: Gardiner's *Fauna and Geography of the Maldive and Laccadive Archipelagoes*, London, vol. 1, 101–18.

PUNNETT, R. C., 1901c. *Lineus. L.M.B.C. Mem. typ. Br. mar. Pl. Anim.*, no. 7, 1–37.

PUNNETT, R. C., 1903. On the nemerteans of Norway. *Bergens Mus. Årb.*, no. 2, 1–35.

PUNNETT, R. C., 1907. On an arboricolous nemertean from the Seychelles. *Trans. Linn. Soc. Lond.*, Ser. 2, Zool., **12**, 57–62.

PUNNETT, R. C., and COOPER, C. F., 1909. On some nemerteans from the eastern Indian Ocean. *Trans. Linn. Soc. Lond.*, Ser. 2, Zool., **13**, 1–15.

QUATREFAGES, A. DE, 1846. Études sur les types inférieurs de l'embranchement des Annelés. Mémoire sur la famille des Némertiens (Nemertea). *Annls Sci. nat., Zool.*, Ser. 3, **6**, 173–303.

QUATREFAGES, A. DE, 1849. Recherches anatomiques et physiologiques faites pendant un voyage sur les côtes de la Sicile et sur divers points du littoral de la France. Paris, vol. 2, Némertes.

RATHKE, H., 1843. Beiträge zur Fauna Norwegens. *Nova Acta Acad. Caesar. Leop. Carol.*, **12**, 231–7.

RATHKE, J., 1799. Jagttagelser henhörende til Indvoldeormenes og Blöddyrenes Naturhistorie. *Skriv. nat. Selsk. Kjöbenhavn*, **5**, 83–4.

REID, W. M., 1950. Glycogen depletion during starvation in the nemertean, *Micrura leidyi* (Verrill), and its ecological significance. *Biol. Bull. mar. biol. Lab., Woods Hole*, **99**, 469–73.

REINHARDT, H., 1941. Beiträge zur Entwicklungsgeschichte der einheimischen Süsswassernemertine *Prostoma graecense* (Böhmig). *Vjschr. naturf. Ges. Zürich*, **86**, 184–255.

REISINGER, E., 1926. Nemertini. Schnurwürmer. *Biologie Tiere Dtl.*, **17**, 7.1–7.24.

REMANE, A., 1940. Einführung in die zoologische Ökologie der Nord- und Ostsee. In: Grimpe and Wagler, *Die Tierwelt der Nord und Ostsee*, 1a, 1–238.

REMANE, A., 1958. Ökologie des Brackwassers. In: *Die Binnengewässer. Einzeldarstellungen aus der Limnologie und ihren Nachbargebieten*. Vol. 22. Die Biologie des Brackwassers. Ed. A. Thienemann. E. Schweizerbart'sche Verlag. Stuttgart, 1–216.

RENIER, S. A., 1847. Del *Tubulanus polymorphus* (Ren.) e della sottoclasse dei Sifonidi. (Osservazione postume di Zoologica Adriatica del Professore Stefano Andrea Renier). Ed. G. Meneghini, Venice, 57–66.

REUTTER, K., 1967. Untersuchungen zur ungeschlechtlichen Fortpflanzung und zum Regenerationsvermögen von *Lineus sanguineus* Rathke (Nemertini). *Wilhelm Roux Arch. EntwMech. Org.*, **159**, 141–202.

REUTTER, K., 1969a. Biogene Amine im Nervensystem von *Lineus sanguineus* Rathke (Nemertini). *Z. Zellforsch. mikrosk. Anat.*, **94**, 391–406.

REUTTER, K., 1969b. Das Verhalten des aminergen Nervensystems während der Regeneration des Vorderdarms von *Lineus sanguineus* Rathke (Nemertini). *Z. Zellforsch. mikrosk. Anat.*, **102**, 283–92.

REUTTER, K., and SASSE, D., 1970. Chemomorphologische Befunde (Glykogen, G-6-P-DH, RNS) am Regenerationsblastem von *Lineus sanguineus* Rathke (Nemertini). *Wilhelm Roux Arch. EntwMech. Org.*, **165**, 285–95.

RIEDL, R., 1959. Das Vorkommen von Nemertinen in unterseeischen Höhlen. *Pubbl. Staz. zool. Napoli*, **30**, Suppl., 529–90.

RIEPEN, O., 1933. Anatomie und Histologie von *Malacobdella grossa* (Müll.).
Z. wiss. Zool., **143**, 323–496.

RIOJA, E., 1941. Hallazgo en xochimilco de *Stichostemma rubrum* (Leidy),
nemerte de agua dulce. *An. Inst. Biol. Univ. Méx.*, **12**, 663–8.

ROE, P., 1967. Studies on the food and feeding behaviour of the nemertean *Paranemertes peregrina*. M.Sc. thesis, Univ. Washington, 44 pp.

ROE, P., 1970. The nutrition of *Paranemertes peregrina* (Rhynchocoela: Hoplonemertea). I. Studies on food and feeding behaviour. *Biol. Bull. mar. biol.
Lab., Woods Hole*, **139**, 80–91.

ROE, P., 1971. Life history and predator–prey interactions of the nemertean
Paranemertes peregrina Coe. Ph.D. thesis, Univ. Washington, 129 pp.

SALENSKY, W., 1884. Recherches sur le developpement du *Monopora vivipara*
(*Borlasia vivipara* Uljan.). *Archs Biol., Paris*, **5**, 517–71.

SALENSKY, W., 1886. Bau und Metamorphose des Pilidium. *Z. wiss. Zool.*, **43**,
481–511.

SALENSKY, W., 1909. Über die embryonale Entwicklung des *Prosorhochmus
viviparus* Uljanin (*Monopora vivipara*). *Izv. imp. Akad. Nauk.*, Ser. **6**, **3**,
325–40.

SALENSKY, W., 1912. Morphogenetische Studien an Würmern. II. Über die
Morphogenese der Nemertinen. I. Entwicklungsgeschichte der Nemertine im
Inneren des Pilidiums. *Zap. imp. Akad. Nauk*, Ser. 8, **30**, no. 10, 1–74.

SALENSKY, W., 1914. Morphogenetische Studien an Würmern. II. Die Morphogenese der Nemertinen. II. Über die Entwicklungsgeschichte des *Prosorhochmus viviparus*. *Zap. imp. Akad. Nauk*, Ser. 8, **33**, no. 2, 1–36.

SANDOZ, H., 1965. Sur la régénération antérieure chez le Némertien *Tetrastemma
vittatum* (Bürg.). *C. r. hebd. Séanc. Acad. Sci., Paris*, **260**, 4091–2.

SCHARRER, B., 1941. Neurosecretion. III. The cerebral organ of the nemerteans.
J. comp. Neurol., **74**, 109–30.

SCHEPOTIEFF, A., 1912. Über die Bedeutung der Wassermannschen Reaktion
für die biologische Forschung. *Zool. Anz.*, **41**, 49–54.

SCHMIDT, G. A., 1931a. Untersuchungen über die Embryologie der Nemertinen.
2. Die Pilidien von *Cerebratulus pantherinus* und *Cerebratulus marginatus*.
Arch. Zool. Ital., **16**, 831–7.

SCHMIDT, G. A., 1931b. Untersuchungen über die Embryologie der Nemertinen.
I. Der zweite entwicklungstypus bei *Lineus ruber* Müll. von der Murman-
küste. *Arch. Zool. Ital.*, **16**, 821–30.

SCHMIDT, G. A., 1934. Ein zweiter Entwicklungstypus von *Lineus gesserensis-
ruber* O. F. Müll. (Nemertini). *Zool. Jb.*, **58**, 607–60.

SCHMIDT, G. A., 1937. Bau und Entwicklung der Pilidien von *Cerebratulus
pantherinus* und *marginatus* und die Frage der morphologischen Merkmale
der Hauptformen der Pilidien. *Zool. Jb.*, **62**, 423–48.

SCHRÖDER, O., 1913. Beiträge zur Kenntnis von *Geonemertes palaensis* Semper.
Abh. senckenb. naturforsch. Ges., **35**, 155–75.

SCHULTZE, M. S., 1851. *Beiträge zur Naturgeschichte der Turbellarien*. C. A.
Koch, Greifswald, 78 pp.

SCHULTZE, M. S., 1853. Zoologische Skizzen. *Z. wiss. Zool.*, **4**, 178–95.

SMITH, C. C., JACKSON, B., and PROSSER, C. L., 1940. Responses to acetylcholine and cholinesterase content of *Cerebratulus*. *Biol. Bull. mar. biol. Lab.,
Woods Hole*, **79**, 377.

SMITH, J. E., 1935. The early development of the nemertean *Cephalothrix rufi-
frons*. *Q. Jl. microsc. Sci.*, **77**, 335–81.

* SOUTHGATE, A. J., 1957. Checklist of marine nemertines from Antarctica.

R. Soc. New Zealand Antarct. Res. Comm., Spec. Rep. no. 22, 1–10.

SOWERBY, J., 1806. *The British Miscellany.* London, p. 15.

STIASNY-WIJNHOFF, G., 1923. On Brinkmann's system of the Nemertea Enopla and *Siboganemertes weberi*, n.g., n.sp. *Q. Jl. microsc. Sci.*, 67, 627–69.

STIASNY-WIJNHOFF, G., 1925. On a collection of nemerteans from Curaçao. *Bijdr. Dierk.*, 24, 97–120.

STIASNY-WIJNHOFF, G., 1926. The Nemertea Polystylifera of Naples. *Pubbl. Staz. zool. Napoli*, 7, 119–68.

STIASNY-WIJNHOFF, G., 1930. Die Gattung *Oerstedia. Zoöl. Meded., Leiden*, 13, 226–40.

STIASNY-WIJNHOFF, G., 1938. Das Genus *Prostomn* Dugès, eine Gattung von Süsswasser-Nemertinen. *Archs néerl. Zool.*, 3, Suppl., 219–30.

STIASNY-WIJNHOFF, G., 1942. Nemertinen der Westafrikanischen Küste. *Zool. Jb.*, 75, 121–94.

STIMPSON, W., 1855. Descriptions of some of the new marine Invertebrata from the Chinese and Japanese Seas. *Proc. Acad. nat. Sci. Philad.*, 7, 375–84.

TAKAKURA, U., 1897. On a new species of *Malacobdella* (*M. japonica*). *Annotnes zool. jap.*, 1, 105–12.

THOMPSON, C. B., 1900. *Carinoma tremaphoros*, a new mesonemertean species. *Zool. Anz.*, 23, 627–30.

THOMPSON, C. B., 1901. *Zygeupolia litoralis*, a new heteronemertean. *Proc. Acad. nat. Sci. Philad.*, 53, 657–739.

TUCKER, M., 1959. Inhibitory control of regeneration in nemertean worms. *J. Morph.*, 105, 569–99.

UCHIDA, T., YAMADA, M., IWATA, F., OGURO, C., and NAGAO, Z., 1963. The zoological environs of the Akkeshi marine biological station. *Publs Akkeshi mar. biol. Stn*, no. 13, 1–36.

VERNET, G., 1966. Les pigments de *Lineus ruber* (O. F. Müller) (Hétéronémertes Lineidae). *C. r. hebd. Séanc. Acad. Sci., Paris*, 263, 191–4.

VERNET, G., 1968. Sur la conversion de l'acide △-aminolevulinique en porphobilinogène par *Lineus ruber* (Hétéronémertes). *C. r. hebd. Séanc. Acad. Sci., Paris*, 266, 18–20.

VERNET, G., 1970. Ultrastructure des photorécepteurs de *Lineus ruber* (O. F. Müller) (Hétéronémertes Lineidae). I. Ultrastructure de l'oeil normal. *Z. Zellforsch. mikrosk. Anat.*, 104, 494–506.

VERNET, G., and GONTCHAROFF, M., 1971. Différenciation des cellules pigmentaires à Porphyrines chez *Lineus ruber* (O. F. Müller) (Hétéronémertes Lineidae). *Histochemie*, 27, 69–77.

VERRILL, A. E., 1892. The marine nemerteans of New England and adjacent waters. *Trans. Conn. Acad. Arts Sci.*. 8, Pt. 2, 382–456.

VINCKIER, D., DEVAUCHELLE, G., and PRENSIER, G., 1970. *Nosema vivieri* n. sp. (Microsporidae, Nosematidae) hyperparasite d'une Grégarine vivant dans le coelome d'une Némerte. *C. r. hebd. Séanc. Acad. Sci., Paris*, 270, 821–3.

WHEELER, J. F. G., 1934. Nemerteans from the South Atlantic and southern oceans. '*Discovery*' *Rep.*, 9, 217–94.

WHEELER, J. F. G., 1937. Nemertea. *Scient. Rep. John Murray Exped., Zool.*, 4, 79–86.

WHEELER, J. F. G., 1940. Notes on Bermudan nemerteans: *Gorgonorhynchus bermudensis*, sp. n. *Ann. Mag. nat. Hist.*, Ser. 11, 6, 433–8.

WIJNHOFF, G., 1913. Die Gattung *Cephalothrix* und ihre Bedeutung für die Systematik der Nemertinen. II. Systematischer Teil. *Zool. Jb.*, 34, 291–320.

WIJNHOFF, G., 1914. The proboscidian system in nemertines. *Q. Jl. microsc. Sci.*, **60**, 273–312.

WILLIAMS, T., 1852. Report on the British Annelida. *Rep. 21st Mtg Br. Ass.*, *Ipswich, 1851*, 159–272.

WILLMER, E. N., 1970. *Cytology and Evolution*. Academic Press, New York and London, 2nd ed., 649 pp.

WILSON, C. B., 1899. Fission and regeneration in *Cerebratulus*. *Science, N.Y.*, **9**, 365.

WILSON, C. B., 1900. The habits and early development of *Cerebratulus lacteus* (Verrill). *Q. Jl. microsc. Sci.*, **43**, 97–198.

WILSON, E. B., 1903. Experiments on cleavage and localisation in the nemertine egg. *Arch. EntwMech. Org.*, **16**, 411–60.

YAMAOKA, T., 1940. The fauna of Akkeshi Bay. IX. Nemertini. *J. Fac. Sci. Hokkaido Univ.*, Ser. 6, Zool., **7**, 205–63.

* YASHNOV, V. A., 1948. Nemertini. Check list of the fauna and flora of the northern seas of the U.S.S.R. Moscow, 89–91.

YATSU, N., 1904. Experiments on the development of egg fragments in *Cerebratulus*. *Biol. Bull. mar. biol. Lab., Woods Hole*, **6**, 123–36.

YATSU, N., 1910a. Experiments on cleavage in the egg of *Cerebratulus*. *J. Coll. Sci. imp. Univ. Tokyo*, **27**, Art. 10, 1–19.

YATSU, N., 1910b. Experiments on germinal localisation in the egg of *Cerebratulus*. *J. Coll. Sci. imp. Univ. Tokyo*, **27**, Art. 17, 1–37.

ZELENY, C., 1904. Experiments on the localisation of developmental factors in the nemertine egg. *J. exp. Zool.*, **1**, 293–329.

ZHURAVLEVA, N. G., KOROTKEVICH, V. S., and KOROTKOVA, G. P., 1970. Vosstanovitel' nye morfogenezy u nemertin. *Arkh. Anat. Gistol. Embriol.*, **59**, 12–22.

INDEX